Orkney Islands
Skara Brae
Pentland Firth
C000175005
Lerwick
Moray Firth
Covesea
Elgin
erness
Sands of Forvie
ore
Rothiemurchus
rngorm Mountains
Aberdeen
Braemar
kwater
rvoir
The Mounth
ch
Dundee
Perth
Bell Rock
Sherrifmuir
ng
Firth of Forth
Bass Rock
Inchkeith
Edinburgh
hern Uplands
Devil's Beef Tub
way Firth
England
0
50 km
0
50 miles

THE UNREMEMBERED PLACES

THE UNREMEMBERED PLACES

Exploring Scotland's Wild Histories

PATRICK BAKER

First published in 2020 by
Birlinn Limited
West Newington House
10 Newington Road
Edinburgh
EH9 1QS

www.birlinn.co.uk

Copyright © Patrick Baker 2020

The right of Patrick Baker to be identified as Author of this work
has been asserted by him in accordance with the Copyright,
Designs and Patents Act 1988.

All rights reserved. No part of this publication may be reproduced,
stored or transmitted in any form without the express written
permission of the publisher.

ISBN: 978 1 78027 637 3

British Library Cataloguing-in-Publication Data
A catalogue record for this book is available from the British Library.

Papers used by Birlinn are from well-managed forests
and other responsible sources

Typeset by Initial Typesetting Services, Edinburgh
Printed and bound by Clays Ltd, Elcograf S.p.A.

For Rachel and for Andrew

Contents

List of Illustrations ix

The Unwritten Places
The Glen Loin Caves, Succoth, Argyll and Bute 1

Blood and Concrete
The Blackwater Reservoir, Lochaber, West Highlands 11

The Memory Path
Jock's Road, Braemar to Glen Clova 35

Wild Island
Inishail, Loch Awe, Argyll and Bute and *Eilean Fhianain, Loch Shiel, Lochaber, West Highlands* 63

Land of the Left Behind
The Atlantic Wall, Sheriffmuir, Clackmannanshire and *the Ardnamurchan Peninsula* 89

Islands of Industry
The Slate Isles, Firth of Lorn, Inner Hebrides 117

Sea Fortress
Inchkeith, Firth of Forth 147

The Bone Caves
Inchnadamph, Assynt 177

Wood of the Ancient Well
Munlochy, the Black Isle 197

Acknowledgements 205

Bibliography 207

Index 217

List of Illustrations

Glen Loin Caves, Succoth

The navvies' graveyard at Blackwater Dam

The dam-edged western end of Blackwater Reservoir

Jock's Road descending from higher ground into Glen Doll

The tiny Davy's Bourach shelter on the Jock's Road drovers' route

On the beach at Inishail island

The cemetery on Inishail

The remains of the replica of the Atlantic Wall, Sheriffmuir

An illicit still hidden in Ardnamurchan

The Clearance village of Bourblaige

Paddling across the Firth of Lorn towards Belnahua

The flooded quarry on Belnahua

Military ruins on Inchkeith Island, Firth of Forth

The unique geology and landscape of Assynt

The Bone Caves, Assynt, and the Allt nan Uamh stream

The Clootie Well, Munlochy

Air cruas nan creag
tha eagar smuaine,
air lom nam bean
tha 'n rann gun chluaine

On the hardness of rocks
is the ordered thought,
on the bareness of mountains
is the forthright verse

from 'Craobh nan Teud' /
'The Tree of Strings'
by Somhairle MacGill-Eain /
Sorley MacLean

The Unwritten Places

The Glen Loin Caves, Succoth, Argyll and Bute

The words seem conspiratorial: secretive, colluding. I mumble them in a half-whisper, trying again to reference them on my map. It's no good. The directions are too vague, too inscrutable for precise placement. I found them online two months ago, and I've been trying to decode them ever since. But maybe that's the point. They are deliberately equivocal, to deter the undetermined, maintain the elusiveness of the place.

'Once in the glen,' they instruct, 'take the forestry road northwards along the valley floor. When the track ends, carry on. Cross an area of broom and gorse, past a hill stream until you reach a high-voltage pylon. Find the narrow path (you'll know it when you see it) and move uphill for a while. Look hard, and eventually you will spot them, not easily at first, but they are there, hidden amid a landslide of colossal boulders.'

The torch beam bounces across the darkness, finding nothing. We keep walking, crunching out a kilometre or so on the rutted forestry track. Mist settles on my face in cold pinpricks, quickly drenching. From somewhere nearby there's the smell of peat smoke, pungent and sweet, lingering in the nostrils and sharp on the tongue. Clumsily we vault a gate, boots slipping on shiny aluminium, the sound of metal ringing in the blackness. I am with my friend Chris – a willing accomplice in these kinds of ventures. We don't talk much; each of us puzzling the clues in the text, searching for a response in the landscape. Soon we have left the road and drifted into thick forestry: birch trees at first, their springy limbs whipping against us; then tightly packed pines, eye-level branches so sharp and brittle I walk with my hands in front of my face, flinching.

We're lost. Or at least, we have lost our way, turning circles in a dead end of rotten tree stumps and dank undergrowth. But then, from close by, I hear something. The purling notes of falling water. I move towards it, suddenly reinvigorated, clutching the soggy sheet of directions, looking for the hill stream. Instead I find a tiny burn; though it's not even that. At best it's a streamlet, a pathetic gurgle of water sluicing between tree trunks – hardly the way-marker we have been looking for. The ground is soft, I've sunk ankle-deep and my boots are now sodden. I'm tired and I'm thinking of returning in daylight, when Chris motions from a break in the treeline.

I know it's there, even before I reach it. The air is charged, a fizzing hum that I can feel as much as I can hear. The pylon rises in a small clearing of heather and gorse as power lines angle in and out from above the tree-tops. We walk underneath the giant structure, and I'm convinced that I can feel the loose energy intensify; a disturbing thrumming in my gums, and a prickling at my fingertips. We search the perimeters of the open ground and find what we are looking for. Part covered by low branches, we spot the beginnings of the path.

There's no doubt now. The directions are right, and the sight of the path removes any uncertainty: it's narrow and deeply gouged, a time-worn furrow, the result of decades, perhaps even centuries, of discreetly acquainted footfall. We scramble upwards, clutching at rocks and stepping between the knuckles of ancient tree roots. I can sense the passage of others here. The delicate tracery on the forest floor of those who have surreptitiously, knowingly, trodden these slopes before us: climbers, brigands, drovers, outlaws. And now, on this remote hillside, we're searching for the same thing.

Although I cannot say when or how I first became aware of the Glen Loin Caves, it feels like they have always been there, hard-wired into my imagination. Most people will never have heard of them, but I have come to think of them in near-mythical terms: an unconfirmed place, conjured somehow within my consciousness over the years by the slow drip-feed of rumour and folklore. When their name occasionally surfaces, in stories about the early days of climbing or mentioned as the hideout for some dubious historical figure, it always stirs a strange restlessness in me.

There is something inherently beguiling about caves anyway, a powerful sense of attraction and foreboding. To consider entering a cave is to experience a conflict of feelings, a potent and contrary rush of curiosity and trepidation, inquisitiveness and apprehension. Caves are portals, breach points where the surface landscape is pierced and an inner world is reached. They are often retainers of mystery as well as space, an ingression into past happenings as much as they are themselves an ingression into the land.

The Glen Loin Caves are loaded with a similar duality, a peculiar and opposing combination of significance and secrecy. On the one hand is their historical importance. Among other claims, they were the reputed resting point for Robert the Bruce and his routed army in 1306 after his defeat at the Battle of Methven. More recently, the maze of fallen rocks on this Argyll mountainside was the focal point of a unique, sporting counterculture.

It was here for almost two decades from the 1920s that groups of working-class young people, mainly from the poverty-stricken tenements of Glasgow and shipyards of Clydebank, congregated to climb the huge rock walls of the Arrochar Alps. They created an almost permanent weekend residence in the caves. Small groups arrived at first, each with its own particular rules and hierarchies, then more established affiliations evolved. Clubs formed here whose names still resonate with modern mountaineers: the Ptarmigan Club and the infamous Creagh Dhu. The influence of these pioneering climbers was immense, providing a surge in climbing standards and techniques that was unequalled anywhere else at the time. They also redefined the sport, dismantling existing class barriers and creating a makeshift society in the Glen Loin Caves whose values and ethics became imprinted on generations of climbers that were to follow.

Yet the caves and their whereabouts have managed to remain largely unknown for decades. Hidden partly by the obscurity of the landscape, but also by an unwritten code of fraternal discretion. 'The lad with the clinker-nailed boots and the rope in his rucksack who told me how to find the cave made me promise to keep the secret,' wrote Alastair Borthwick in 1939, in one of the earliest and most detailed descriptions of the caves. 'I was to follow a track to a forester's cottage, pass through a gate . . . and there search for an old sheep fank. Behind it I should find a faint track leading up the hillside; and if I followed the scratches on the rock it led to, I should find the cave and good company.'

Even at close proximity, however, the caves are frustratingly hard to locate. In 1996 the writer Rennie McOwan

The Guardian Bookshop

A BONE
11A CASTLEGREEN STREET
DUMBARTON
WEST DUNBARTONSHIRE

G82 1HN
UNITED KINGDOM

THE GUARDIAN BOOKSHOP
PO BOX 48
WESTHAM, EAST SUSSEX
BN23 6WB

Thank you for ordering via
The Guardian Bookshop
We do hope you enjoy your product
If there are any issues please contact us
on 0203 176 3837 or
help@guardianbookshop.com

Order ID: [4000092943]

Order date: [06/10/2020]

UPC	TITLE	QTY
9781780276373	UNREMEMBERED PLACES	1

Order 4000092943

We hope you and your family are well and safe

described his efforts to find them. 'This huge tangle of steep rocks, high up a hillside in Glen Loin . . . is not easily found. The ground is rough, very steep, often cliff-like and a mass of tree-covered holes, fissures and crevices.' It took McOwan, an accomplished outdoorsman, several attempts to pinpoint the exact position of Borthwick's earlier description, leaving no doubt about the visual indiscernibility of the caves. 'You can trace these historic caves if you know where to look', McOwan advised matter-of-factly, 'but it can be both time-consuming, and exasperating if you do not.'

I am captivated by these kinds of places. Although compelled is perhaps a better word to describe the slightly obsessive nature of my interest. For many years I had regularly roamed Britain's largest and most inhospitable mountain range – the Cairngorms – searching for something akin to what the writer Roger Deakin had described as 'the unwritten places': fragments of human and natural history that had somehow become lost in that vast granite landscape of plateau and corrie. Like the Glen Loin Caves, these were peripheral places, existing at the edges of our collective memory and often hidden by dint of sheer geographical remoteness. I had come to think of them best described as wild histories. Wild, certainly, in that they were located in wilderness areas, but wild also in an almost anthropomorphic sense: feral, uncared for, mostly unknown or nameless, and outside the boundaries of public consciousness.

It is hard to believe that in such a densely populated archipelago as ours there are features of our landscape that could remain undocumented or unexplained, that there are places beyond our comprehension or recollection. Perhaps this is because we have become disconnected – distanced both physically and in thought – from the familiarities of wild places. So much so that we have come to regard our history with a distinctly contemporary, geographical bias: a predominantly categorised, class-bound and urban interpretation of the past. But this is forgetting that we have only relatively recently become a nation of city-dwellers, and that Britain's northern latitudes are still a place of wildness, a littoral-edged domain, full of mountain and moor, forest and fen. And it is from these places that we have ancestrally travelled.

When thought of in this way, the landscape of Scotland becomes a vast diorama: the setting for countless narrative scenes, lives and stories overlaid, some more vivid than others. These wild histories define us, perhaps more than any iconic building or national monument, for they are records of things inconsequential and commonplace. They are the simple transactions of life and land, of life *in* land. The same repetitive priorities that echo distantly in our own lives today.

Our islands are deep in time, but limited in their boundaries, and are therefore densely layered in mystery and significance. Anyone who has spent time in Scotland's more remote regions or has purposely explored its less-visited nooks (and crannogs) may well have come across some fragment of a recent or a long-forgotten past – for wild histories are profuse here, often hidden in plain sight but invariably difficult to reach. They are the strange anomalies in the landscape encountered by chance on an isolated ridgeline or discovered on a stretch of deserted coast. They appear without explanation or ceremony, harbouring stories of uncertain origin: apocryphal tales with a hint of truth, enough to seed intrigue or perpetuate a myth.

It would be impossible to search for or catalogue all of Scotland's wild histories. To do so would involve a lifetime's exploration and would, by the passage of time, be rendered incomplete even before it was finished. But I wanted to reach certain places which, through their location and mysteriousness, had for years exerted on me a powerful imaginary appeal. They were often sacred but unremembered sites, such as medieval burial grounds, hidden on remote Highland lochs or the abandoned graveyard for itinerant construction workers of the Blackwater Dam – perhaps the most desolate cemetery in the whole of Britain. There were also curiosities: the chance to visit one of Scotland's highest (and smallest) mountain shelters, situated – if I could find it – somewhere on an ancient drovers' route, as well as the derelict sea island once used as a prison, quarantine site and military garrison, which still guarded the wind-strafed waters of the Firth of Forth. In the Inner Hebrides, I intended to spend a night on Belnahua, one of the uninhabited Slate Isles, where a ghostly village stood watch over the deep lagoons of abandoned slate quarries, flooded by Atlantic storm surges. Underground places would also feature in my journeys, and I would travel to Assynt's karstic landscape in search of the enigmatic Bone Caves. Elsewhere, I would track across empty moorlands looking for the remains of illicit stills and the clues to a secretive bootlegging past.

The journeys would be neither definitive nor conclusive. Neither would they be a search for the unsurpassed: the most 'wild', the most 'remote' or the most 'obscure'. Instead, they would be more folly than analysis, personal rather than primary discoveries. By necessity, they would also only be possible to experience first-hand, by self-made journeys on foot or by boat, and because of this they would also be an exploration of the landscape itself and the forgotten links between people and place.

We're not having much luck. Chris lowers himself into another opening in the rocks – the third we've tried. I stand over the gap and peer in from above, seeing his head torch sweep the interior. The light disappears. I hear shuffling and some words I can't make out, followed by silence – then a call from lower down on the other side of the rock. 'No good.' Chris emerges from a vegetated crack in the hillside below. 'Too small, too damp. That can't be it.'

By now the rain has stopped and the cloud cover has thinned. I can see clusters of stars through breaks in the forest's canopy. It's close to freezing and my breath lingers in the thin cone of torch light. We continue higher, zig-zagging steeply through pines, tracking a chute of massive boulders. I have the feeling we are getting closer, but I'm being careful, remembering another description I have read about the area. The mountaineer Hamish Brown had struck a cautionary tone. The place, he warned, is 'riddled with caves and howffs of all sizes. Some overgrown gashes can provide booby-traps every bit as dangerous as crevasses'.

The gradient eases and then I see it. Ahead of me is a curtain of rock, glossy and bright in the moonlight. But there's something else – a thin pleat of darkness. I move closer, and as the angle changes the crimp becomes a wide triangle, a large void of textureless black. The opening is huge: a story-book cave entrance, an eight-foot-high archway with tendrils of gnarled tree roots snaking along the threshold. It's so perfectly formed it could be straight from a fairy tale: a bear's den, the home to an ogre or a band of thieves. We enter slowly – in real life, caves can still be places for those not wanting to be found.

It's dry inside, despite the rain. The ground is dusty and strewn with boulders. Sound redoubles, each movement carrying a louder, secondary reverberation. It feels like we have entered a crypt. A large, cold space: vaulted and full of dark air. I scan the cave walls, seeing ripples and folds appear in the torchlight, waves curving and bending in the schist, ridged to the touch – a kind of metamorphic graffiti. It's laughable, but I'm ridiculously pleased to have found the cave, finally closing a loop of such long-standing fixation. More immediately, though, as the temperature plummets, it also means we have shelter for the night. Chris sets about making camp, arranging his sleeping bag between the rocks. From his jacket he has unstowed a plastic bottle with whisky swilling inside, straw-coloured and gleaming.

I take off my pack and explore further in. There's an anteroom, a narrower chamber that I clamber into. It leads back into the open and I find myself at the bottom of a small chasm with rock walls rising either side of me. Water falls in thick, rhythmic droplets from the branches above. I work my way along the fissure, wading through slippery rock pools and pressing my hands sideways to balance, my fingers sinking into sponges of damp moss.

My route is soon blocked by a steep ramp of boulders. About halfway up I see another large cavity, hard to reach in the wet conditions without climbing gear. Borthwick had described finding something similar – 'holes' which appeared ominously 'to lead directly into the bowels of the earth', and I wonder if this is the same huge cave 'about forty feet square with a roof fifteen feet high', that he had discovered.

Borthwick told of a boisterous place, noisy but welcoming, where 'someone was always arriving' – the cave being home to a rowdy and garrulous lot: 'As the shouting grew, others arrived. We had eighteen in residence in the end . . . Then they told stories . . . They seemed to have been in every conceivable variety of scrape on every conceivable variety of mountain, and the bigger the scrape the louder the laughter.'

I picture the scene as if I were arriving many decades ago. Not much would have been different; the same uncertain, perilous route to get here. But there, in the cliff face high above me would be the cave's fire-lit entrance: a hot coal, bright and singular in the darkness, with loud voices barrelling out into the night.

Blood and Concrete

The Blackwater Reservoir, Lochaber, West Highlands

At the start of the twentieth century, in a remote glen in the West Highlands, the clatter of pickaxes and voices rings out: hard metallic sounds and the great compound noise of human commotion. A dam is being built, an alien shape in the landscape, linear and distinct, cleaved into the steep undulations of hillside.

The place hives with activity. Rock and peat are blasted away, and smoke blows through the cranes and rigging, billowing past small shanty-town huts and out across moorland. Thousands of men are at work, an army of the desperate and dispossessed. There's blood and toil to be found in the mud and heather here, and hardship and death.

'There was a graveyard in the place . . .' wrote Patrick MacGill about the building of Blackwater Reservoir in his thinly veiled autobiographical novel *Children of the Dead End*. 'A few went there from the last shift with the red muck still on their trousers and their long unshaven beards still on their faces. Maybe they died under a fallen rock or a broken derrick jib. Once dead they were buried, and there was an end of them.'

MacGill's book is one of the most brutal I have ever read. It tells the story of Dermod Flynn, a feisty adolescent forced from his home in Ireland into bonded labour in Scotland. Years of itinerant work and unremitting poverty eventually lead him to Kinlochleven. It is here, along with so many other Irish and Scottish navvies, that he finds employment in the hydroelectric scheme: a massive civil-engineering project which included the construction of the reservoir, a six-kilometre aqueduct and an aluminium-smelting plant.

The chapters describing Flynn's (or rather, MacGill's) life at Kinlochleven are among the most powerful and disturbing in the novel. The squalor of the workers' encampment, where the 'muddle of shacks' looked as though they had 'dropped out of the sky' and out of which 'a spring oozed through the earthen floor', is only matched by the danger of the tasks they are required to carry out: 'As he struck the ground there was a deadly roar; the pick whirled around, sprung upwards, twirled in the air like a wind-swept straw, and entered Bill's throat just a finger's breadth below the Adam's apple. One of the dynamite charges had failed to explode on the previous day, and Bill had struck it with the point of the pick, and with this tool which had earned him his livelihood for many years sticking in his throat he stood for a moment swaying unsteadily. He laughed awkwardly as if ashamed of what had happened, then dropped silently to the ground.'

Children of the Dead End is deeply affecting: in equal measure hard to read and hard to forget. It reverberated deep in my subconscious long after the last page was finished with the kind of protracted background hum that all potent books seem to leave behind. The characters and their place within the Highland landscape were like nothing else that I had ever come across. This was no idealised version of nature, no celebration of a wild, but ultimately beautiful place. Neither was it a setting that prompted any prospect of spiritual reclamation, nor transcendental enlightenment for the navvies. The place was pitiless and unforgiving.

MacGill describes an almost dystopian society, a far-flung outpost of lawlessness where 'all manner of quarrels were settled with fists', and drinking and gambling are the only possible distractions from the savagery of working life. The mountains, the moor, the sheer scale and grandeur of scenery are no solace. No succour is to be found in the landscape's aesthetics. Instead, the environment is merely another agent of misery for the men. The navvies are thus caught between two contrasting but overlapping adversities, asperities to be endured that are both wild and man-made, natural and industrial.

It's not just the matter-of-fact wretchedness of the Kinlochleven descriptions, though, that render MacGill's book extraordinary. More significantly, the prose, in all its awkward mixture of autobiography-cum-fiction, is an otherwise untold story. It becomes the counterpoint legacy to the mass of concrete and steel, a parallel and forgotten voice to be measured against the cold, physical reality of the Blackwater Dam. MacGill's text provides a human narrative, the collective testament for the thousands of men who were once part of this wild history.

Of course, many never left the place. For some, all the inherent danger of the work would coalesce in a single instant. With a sudden evaporation of luck – the misplaced sledgehammer blow, a moment's loss of balance or the abrupt death-strike of unseen rock-fall – lives were ended in the wind-torn reaches of the moor. They were laid to rest near where they fell, in a small, improvised burial ground situated below the steep walls of the reservoir. This was the place MacGill had described in *Children of the Dead End* – the navvies' graveyard. Over a century after the book was written, it retains a strange literal and literary identity, a gangplank extending between fiction and reality, existing both on the pages of MacGill's novel and on the empty moorland: a handful of weather-beaten concrete headstones in a scarce, vacated landscape.

It was my fourth attempt to reach the place. A couple of years earlier I had made the journey north after work, my eyes gritty and screen-burnt from hours of working at my computer. That autumn night I had driven into the remnants of a tropical cyclone. Hurricane Gonzalo, the most destructive Atlantic storm to occur in several years, had spun its way out across the ocean and was making landfall again, a condensed set of isobars and weather fronts hitting the west coast of Scotland in a maelstrom of gales and heavy rain.

In some idiotic form of logic, I was hoping for a weather window. If I were to go directly into the weather, I had concluded, I would – at some point on my journey – encounter the storm's eye, and thereby be graced with several hours of benign, tranquil conditions.

Forestry lorries thundered past me, orange lights flashing, washing up walls of spray and surface water. For several hours I assumed a white-knuckle driving position, hunched forward on the steering wheel, squinting through the windscreen, the wipers knocking out a frantic rhythm. I eventually gave up. Fraught and tired, I pulled off the road somewhere before Rannoch Moor, pitching my tent on rain-soaked ground, the fly sheet snapping violently in the wind.

I tried twice more: beaten back before I had even started by thick winter snowfall; then, in spring, halted after only a couple of hours of walking, a back sprain leaving me almost immobile five kilometres into the disturbingly empty moorland. This time, I had subconsciously decided, I would definitely get there.

Conditions were good as I reached the Devil's Staircase. High-altitude clouds moved in slow south-westerly convoys, the sun behind them, chalky and bright. The five-hundred-metre-high pass rose ahead of me in a switchback of loose grit and polished rock. Despite the supernatural association, it looked innocuous enough, hardly befitting such a foreboding name.

Similar toponymic curiosities occur frequently in the British Isles. There are over eighty places that Satan has laid a proprietary claim to across Britain. Often, they are some of the country's most spectacular and unusual landforms, from fragile sea arches and precipitous gorges to strange rock formations and large cave structures. The names are equally memorable and explanatory; I knew of several first-hand. There was the Devil's Beef Tub in the Scottish Borders, a deep hollow at the intersection of several hills plunging from the moorland above, and the silver-grey granite tor that forms the Devil's Chair in Shropshire.

Most familiar to me was the Devil's Point, the huge sentry-mountain guarding the entrance to the Lairig Ghru pass in the Cairngorms. Its notoriety comes not from the name itself but how the naming came about. As the story goes, the mountain was diplomatically rechristened by a royal attendant. While visiting the area, Queen Victoria enquired after the Gaelic translation of the mountain. 'The Devil's Point', replied a quick-thinking ghillie keen to spare the monarch's (and his own) blushes – a euphemism hastily stumbled upon to conceal the peak's actual name, the Devil's Penis.

The Devil's Staircase was, by comparison, quite unremarkable: appearing neither topographically conspicuous nor malevolent. My contemporary interpretation, though, was skewed. Many years ago this had been a place of ultimatum – the lowest section, and only crossing point, on the spine of hills that separates the vast expanse of Rannoch Moor to the south with the back country above Kinlochleven to the north. In bad weather, a fateful calculation would often have needed to be made – negotiate the pass and save a day of travel, but run the risk of being caught in lethal winter conditions.

Such a deceptive landscape proved fatal for the navvies. 'They used to climb over the back glen there, down to [the] Kingshouse Inn,' wrote Borthwick, retelling the story of a local stalker. 'And then they would try and get back over the glen again, in the dark, and the snow. I used to go up with the pony in the [s]pring, when the snow melted. I've brought down as many as twenty. Poor devils. You'd see maybe a boot or an arm sticking out of a drift . . . I came on a skeleton . . . Thirty year it had been there. There was moss on the bones and a bottle in its hand.'

I knew the stalker's tale was probably tinged with the theatrical, just as Borthwick's account may have extended the truth even further. Nevertheless, it welled a sad admiration in me. The desperation of men willing to commit to a round trip of at least fifteen kilometres, through bog, tussock, snow and darkness for the 'hard stuff' piqued both my sympathy and my respect. From the top of the pass, I could see some of the distance the navvies would have walked, through the narrow breach in the hills, on to the tawny moorland, and, beyond that, the cliff-armoured north-east face of the Buachaille Etive Mor. The Kingshouse was a further five kilometres away, tucked out of sight behind the flanks of another mountain, out there somewhere on the vast moor.

Leaving the pass's summit, I cut across folds of empty moorland, stumbling mainly. For an hour I fell in and out of contour lines, the rhythm of my footfall constantly disrupted by the uncertainty of the terrain: hidden burns, spine-jarring rocks and mossy sinkholes. Eventually I passed the point of my previous attempt and reached a stretch of light-green woodland that flickered in the sunlight and breeze. It was here that I hoped to locate a strange-looking vehicle track that was marked on the map. Something about its description didn't seem right. Its sudden height gain, its elaborate twists and turns, and its elevated position above the glen hinted at some kind of eccentric folly, a mad man's foolish endeavour.

What I found was completely unexpected. I dropped down through a shrubby incline, brushing my way through a thicket of birch and heather, until, without warning, my feet landed solidly on concrete. I was in a lateral clearing, with trees parted either side of me, left and right. I stood on what looked like a walkway, the surface a metre or so wide, perfectly flat and laid in large rectangular sections spanning ahead like tightly packed railway sleepers. Below it, the land continued to fall away so steeply that the structure became a viewing platform onto the glen below.

I moved gingerly on the concrete, uncertain it would hold my weight. Each step rang with a hollow echo. After a few metres, I noticed several bore holes an inch or so wide, spaced at intervals on the surface. They revealed not only the thickness of the concrete, but something much more remarkable underneath – water! Silver and black and moving at incredible speed. This was no walkway. I was on the aqueduct, inches above a man-made torrent of incredible power, the sight of which left me momentarily rooted to the spot.

MacGill's characters in *Children of the Dead End* are part of the unseen: a soluble, temporary population that, at the time of the book's writing in 1914 until roughly the 1950s, existed across Britain in a state of permanent migratory anonymity, uncountable and unaccounted for.

'I was not the only one on the road,' observed Laurie Lee in his wayfaring classic *As I Walked Out One Midsummer Morning*. 'I soon noticed there were many others, all trudging in a sombre procession.' They went by many (often pejorative) names – tramps, journeymen, vagrants, vagabonds, tinkers – drifting through the countryside from city to city, and from job to job in what the historian E.J. Hobsbawm ironically described as 'the artisan's equivalent of the grand tour', their journeying both essential and often purposeless, driven either by economic necessity or simply by a life of habitual wandering.

Little is known about these people either collectively or individually. Yet their presence in Britain's landscape was once ubiquitous; a continuous movement of human traffic that ebbed and flowed across the decades, through winters and summers, in fields and villages, roadsides and towns. They were often reviled and demonised, but accepted somehow as part of the natural order of things. This unconsidered part of society has now become an equally unconsidered part of history. Shadow figures: shambling regiments consigned to memory and old photographs, silently padding along lanes and byways, bedraggled images fading into the landscape.

Britain's itinerant past, then, is defined by transit and transience. A drifting, unanchored way of life which, although long forgotten in its habits and customs, has left a resounding echo in literature. In writing, as in life, the unrooted, the homeless, the wanderers are often portrayed as outsiders, peripheral beings beyond understanding and therefore prefixed with symbolism or mystery.

In Thomas Hardy's *The Return of the Native*, the most enigmatic character is Diggory Venn, a travelling 'reddleman' who wends his way aloofly across Egdon Heath. Venn sells reddle, a dye used for marking sheep, which stains his clothes and skin red, turning him into a surreal 'blood-coloured figure'. The reddleman has an almost supernatural significance in the novel: visually demonic and displaying an uncanny luck at gambling, he is an interloper of other-worldly associations, a loner whose influence is subtle but constant, observing and interceding from a self-imposed distance. Likewise, Samuel Beckett's tramps in *Waiting for Godot* are loaded with similar metaphoric power. Stripped of identity and place, stranded somewhere far apart from society without knowing why they are trapped, imprisoned by both their circumstances and their hope.

It was George Orwell, the intellectual champion of the downtrodden and marginalised, who was the writer perhaps most influenced by what he described as 'the tribe of men, tens of thousands . . . marching up and down England'. Between 1928 and 1931 Orwell regularly 'went native', dressing in ragged clothes and frequenting workhouses in order to research his subject. The experiences were formative (providing the basis of *Down and Out in Paris and*

London), but contrived. The Old Etonian and ex-policeman was always an outsider looking in, and must have appeared desperately out of place. 'I dared not speak to anyone,' confessed Orwell's narrator, 'imagining that they must notice a disparity between my accent and my clothes.'

Few writers could claim the credentials of true vagabondia. Robert Louis Stevenson's verse 'The Vagabond' waxed lyrical about a life (very different from his own middle-class existence), with a 'bed in the bush with stars to see'. And the Edwardian writer Stephen Graham confusingly extolled the virtues of wayfaring in his book in *The Gentle Art of Tramping*, while at the same time expressing open contempt at those who did so without choice. 'They learn little on their wanderings,' declared Graham acerbically, 'beyond how to cadge, how to steal, how to avoid dogs and the police. They are not pilgrims but outlaws.'

By contrast, authenticity came from harsh experience. Jack London, a veteran of multiple hardships, wrote *The People of the Abyss*, having stayed for months in the slums of East London, often sleeping rough on the streets. Few people could have been more qualified to write about the conditions he witnessed. By the age of twenty-one, London had survived a tough childhood, provided for his family by working twenty-hour shifts in a cannery, scraped a living poaching oysters in San Francisco Bay, sailed the Pacific, lived as a tramp (and had been imprisoned for vagrancy) and had prospected for gold in the deathly grip of the Yukon winter.

The Welsh poet and author of *Autobiography of a Super-Tramp*, W. H. Davies, shared a similar pedigree of hoboism, and the physical distresses to prove it. While attempting to jump a freight train, Davies slipped. His foot was crushed under the carriage's wheels, resulting in an injury so severe his lower leg was later amputated.

The literature of the open road is therefore both diverse and revelatory. It has become, by absence of formal record, a prism, refracting and dispersing an unacknowledged part of our social history: a medium of transfer and reflection for lives and stories otherwise untold. But of all the books, journals and accounts that I have come across in this sprawling genre, *Children of the Dead End* is perhaps the most articulate and the most truthful. 'Most of my story is autobiographical,' wrote MacGill. 'I have endeavoured to tell of the navvy; the life he leads, the dangers he dares, and the death he often dies.'

At the time when the novel was first published, in 1914, most written works largely ignored the working classes, either in their subject matter or intended readership. Proletarian literature had yet to fully emerge as an established literary form, and any references to the working classes (even when well-intended) were always made by those of a more educated social standing. MacGill's book, though, was radically different, and instantly successful because of it, selling 15,000 copies in its first three months.

Although a work of fiction, its content was mined from abysmal first-hand experience: 'it must be said that nearly all the incidents of the book have come under the observation of the writer,' declared MacGill, and 'that such incidents should take place makes the tragedy of the story'. Knowing this, and reading the first-person narrative, it is impossible at times to separate author and protagonist.

We follow Flynn/MacGill through a series of adversities. As a child, alone and on the road, he trudges through winter nights to keep warm, sleeps in hedgerows and ditches, and is both threatened and pitied by the adults he encounters. Descriptions are precise and sensory, proximate and compelling. The feeling of water 'gurgling' in his leaking boot, the sound of wind as it passes through telegraph wires, or the sensation of being 'close to the earth, almost part of it', with 'the smell of the wet sod . . . heavy in my nostrils'. Details given are also appallingly graphic. In one particularly gruesome passage, MacGill recounts the death of a railway worker caught beneath a ballast train, the 'soft, slippery movement of that monstrous wheel skidding in flesh and blood', and the sombre act of his colleagues to retrieve the body, finding 'scraps of clothing and buttons' that were 'scrambled up with the flesh'.

At the time of its publication, some critics questioned the gritty realism of the *Children of the Dead End*, not believing that the events in the book and the environment in which they took place could actually exist. But this was because MacGill's prose, honest and unadorned, was so ground-breaking and genuine. Writers who had previously delved into this world – Charles Dickens, Victor Hugo, Émile Zola, Mark Twain – were just that, writers in a realm that was not their own. For MacGill, writing served a different purpose: part catharsis, part social commentary, a means of escape and a deep emanation of personal experience.

Writing saves Flynn, as in reality it did for MacGill. It is an unexpected salvation, bordering on an epiphany and occurring the moment he idly picks up a scrap of paper with verses of a Robert Browning poem handwritten across it. The young man is instantly captivated, enthralled with the poem and with words that 'expressed thoughts of my own, thoughts lying so deeply that I was not able to explain or express them'. From then on, he reads voraciously, obtaining a card for the Carnegie Library and spending all his spare cash on bundles of second-hand books. At work, in Glasgow's railway tunnels, he reads at every opportunity: Carlyle's *Sartor Resartus* while seated on the footplate of a train, Ruskin's *Sesame and Lilies* under a coal wagon and Hugo's *Les Misérables* by the dim light of a naphtha lamp.

He begins penning verses of his own, which he self-consciously reads to his workmates. They like them, and he is spurred to write more. Later, as he journeys to Kinlochleven, he becomes 'possessed of a leaning towards lilting rhymes', composing poems while padding out the long distances to the dam. Once there, he feels 'compelled' to write more emphatically than ever, and after witnessing a fatal accident he puts pen to paper, for no reason other than to 'scribble down the thoughts which entered my mind'. On a whim, he sends the account of the tragedy to a newspaper in London, which publishes it and commissions more articles. This is the pivotal point in the novel, the moment where Flynn's lot finally improves and he begins to unshackle himself from the miserable circumstances of his existence.

In real life, this also marked a turning point for MacGill. He left the life of the navvy and the road behind him, and moved to London to become a journalist. In the years to follow he wrote prolifically: novels, poetry, plays and even screenwriting. He married a fellow Irish writer, Margaret Gibbons, with whom he had three daughters and later moved to America. For the uneducated child labourer from Donegal, forced from home into unthinkable poverty, the transformation was profound and remarkable – a feat of incredible autodidactic prowess, but, perhaps more significantly, even greater human resilience. 'The love of life was strong in me,' wrote MacGill as Flynn, 'and never did I cling closer to life than I did at that moment when it was blackest.'

I soon fell into an easy rhythm, my strides on the aqueduct accompanied by their own booming sound effects. It took me a couple of kilometres to adjust to the theatrical acoustics. For a while, I walked toes first, sheepishly trying to soften my footfall, oddly embarrassed by the noise I was making.

The aqueduct was a structure of altering distinctions. At times it was brash and overt. Hard lines were set straight against the landscape, visible from miles around. At others, it was hidden and secretive, wending through birch copses, tucked neatly to the contours of the hillside, bending and curving in sympathy with the map's re-entrants. I battled with a grudging admiration, disapproving of its intrusion into the wildness, but inwardly impressed at the effort and ingenuity that must have been involved in its construction.

Where the land dropped away sharply, which it often did, bridges had been built. Solid stone columns and archways spanning drops of thirty feet or more lifted the aqueduct over churning white water and the tops of trees. I stepped carefully across each bridge, perturbed by the warning signs: 'Danger. Persons using this structure do so at their own risk'.

At various points, the hillside above had been washed away, crumbling onto the surface of the aqueduct, leaving scatterings of rocks and earth. Where this process had been more sustained and long-term, the concrete had become turfed over with mosses and grass, so that the sounds of my movement were momentarily muffled. In these sections, the aqueduct had felt dark and submerged, grafted back into the mountains. Tiny waterfalls filtered through clusters of green-black ferns so dense that they resembled miniature rainforests.

The weather began to shift as I neared the graveyard. The sky darkened, and to the west rain clouds were forming, fresh from the Atlantic. The land also changed. The ground around me levelled out and I entered into an area of moorland with parallel mountains to the north and south. I knew I was getting close. I passed by an oval-shaped bulge in the river marked on my map as the Dubh Lochan. It held a small island bristling with bright trees: aspen, birch, alder. Ahead of me, I could make out the dam, the flatness of its upper wall pressed neatly against the horizon, a kilometre long. I clambered down from the aqueduct and cut north into a wide swathe of heather.

Within minutes I saw it, bleakly visible in the middle of the moor: a neat rectangle of land surrounding a small tortoiseshell knoll, sectioned off by a low picket fence weathered to bone-grey. Inside there were over a dozen headstones, set in three rows rising up and over the crest of the knoll. Even at a distance, its appearance was surreal. It had a strangely filmic quality: dramatic and out of place, like a set design ready to be used in a High Plains western.

I followed a faint path that had been worn through the heather and stood at the edge of the fence. The gravestones were low in the ground, blotched with lichen and fissured with moss. Each had been made with concrete from the dam, thin rectangles with the top corners angled off and listing at varying angles in the soil. A few had been etched with simple decorative borders, but most just carried names – all either Irish or Scottish. One in particular caught my eye, the grave of a woman, buried alongside the men. I was surprised – my perception of the place suddenly altered. In all the accounts I had read and pictures I had seen, there had never been any sign of women at the dam. It was, I had wrongly assumed, an irrefutably male domain, an ungovernable and wild work camp. But the thought of both men and women working here suggested something more civilised, a settlement or community.

Despite the lack of embellishments, the inscriptions were exact: neat letterings carved with considerable care. The same effort had been applied to all the stones, including the one nearest to me, an anonymous grave which, in the absence of a name, simply had the words 'not known' traced delicately across its centre. The gesture was touching. The stonemasons' precise work, their attention to detail and unbiased craftsmanship, even for an unidentified colleague, felt like an act of considerable compassion.

I stood for some time with the moor's breezes whipping across my face, feeling slightly stupefied by the strangeness of the place, but also wondering at the graveyard's positioning. All around me was flatter, more suitable land, but, for some reason, the workers had chosen to bury their dead on this rocky mound.

Perhaps it was simple practicality. This area of ground, in the shadow of the dam, would have been choked with machinery, materials and the workers' makeshift shelters. 'A sleepy hollow lay below,' wrote MacGill about the scene here, 'and within it a muddle of shacks roofed with tarred canvas, and built of driven piles, were huddled together in bewildering confusion. These were surrounded by puddles, heaps of disused wood, tins, bottles, and all manner of discarded rubbish.' Usable space would have been at a premium, and the awkward knoll would, no doubt, have been difficult to assign to another purpose.

But there was also an alternative reason for the graveyard's placement, something more deliberate and considered. The direction of the headstones seemed significant. They faced east, looking directly out onto the dam, imposing a connection between the workers and their task. As if to say, this is why you came here, and this is why you remain. It had created a strange synchronism in the landscape. Less than a kilometre apart, both the graveyard and the dam had become irrevocably linked with each other, each the reason for the other's being. The use of the knoll's higher ground also appeared meaningful, a conscious choice. It had ensured elevation, a literal and spiritual prominence for the interred. It meant that despite all the lawlessness, the violence and the squalor of the place, a thoughtfulness and humanity had somehow prevailed here.

It was time for me to leave. Gusts began arriving in sets, funnelling through the glen ahead of the approaching storm. The afternoon was drawing out, and I needed to find somewhere to stay for the night. Besides, the solemnness and solitude of the place had begun to unsettle me. I moved away, trying to dislodge the feeling, unaware at that moment of how much my nerves would be tested later that night.

In 1886 two men working on either side of the Atlantic made a discovery that would eventually alter large areas of Highland landscape. It was a breakthrough which ended a metallurgic conundrum that had existed for most of the century, and laid the foundation for the large-scale production of one of the world's most coveted metals.

Aluminium, the most abundant metallic element in the Earth's crust, had only been discovered sixty years earlier by the Danish physicist Hans Christian Oersted. Until then, its existence was merely theorised: appearing naturally in chemical compounds, it had never been found anywhere in the world in its metallic form. But by heating aluminium chloride with potassium amalgam in 1825, Oersted managed a feat of modern alchemy, the first conclusive evidence of the metal – a few small, impure flecks of aluminium.

The mass production of aluminium, though, proved difficult. Despite its geological ubiquity, only minute amounts of the metal could be extracted at a time. Limitations in the metal's supply, combined with its strength and ghostly lightness, meant it was both highly sought after and highly valuable, outstripping the price of gold for a brief period in the mid-nineteenth century.

All that changed though, when Paul Louis-Toussaint Héroult in France and Charles Martin Hall in the United States realised that by a simple process of electrolysis (passing an electric current through a dissolved solution containing aluminium oxide) vast quantities of the metal could be made. Commercial production then became merely a question of scale, of having enough space and enough electricity to enable the smelting process.

With an abundance of uncultivated land and a potentially unlimited supply of hydroelectricity, the Scottish Highlands were a logical setting for this new industry to take root. The British Aluminium Company pioneered developments, first at its metal works near the Falls of Foyers, but then later turning its attention to a remote settlement at the head of

Loch Leven. It was here, beneath darkly sloping hillsides, that two tiny communities, Kinlochbeg and Kinlochmore (totalling two hunting lodges and four cottages), were transformed into a purpose-built industrial village. Semi-detached workers' houses with handsome sash-and-case windows and gable-roofed porches, each with their own small parcels of garden, were laid out in neat, geometric rows. Behind them sat grander homes, larger detached properties for the managers, constructed in the Arts and Crafts style, with lofty entrances and sweeping driveways. Roads were created, lined with telegraph poles and electric street lights. And at the far end of the village sat the huge smelting plant, the Héroult/Hall reduction process enlarged on a massive scale.

By 1908 the village, built from scratch, was complete. 'Aluminiumville' was first thought of as a name, but then rejected. Instead it became Kinlochleven, a beacon of modernity nestled among the empty glens, a model village built by the British Aluminium Company with echoes of Josiah Wedgwood's Etruria and Robert Owen's New Lanark, and all the paternalistic idealism that came with it.

A local newspaper, *The Oban Times*, would later describe the village with effusive language as, 'perhaps the most remarkable and romantic example of Highland, or even of Scottish, industrial development'. The article continued in gushing prose. 'Kinlochleven by night bears an impressive scenic aspect, which makes it worth the visitor's while to tarry until the shades of evening fall. Hosts of brilliant electric lights, every home illuminated, and the twinkling aluminium furnaces in a sombre setting of frowning bens and darkened sea, raise up the fancies of fairyland and the magical power of Pluto.'

Reality, though, was somewhat different. Enclosed by mountains, and subject to high levels of rainfall, the village could endure months without sunlight. Access in and out of Kinlochleven was also challenging. It wasn't until 1922 that a road was built that connected the village to the neighbouring settlement of North Ballachulish. Until then, the journey to and from the village could only be made by boat or boot, compounding the sense of claustrophobia the residents must have felt. For them, day-to-day routines began and ended in the geographical and social confines of Kinlochleven's careful town planning. Home life and work life were indivisible, dominated by the omnipresent smelter, its dark silhouette visible from every part of the village, a fug of smoke and fumes above. Then there was the noise, the constant drone of the plant machinery always audible, the maddening whine and hum of progress.

But even as the British Aluminium employees began to colonise the new village, fulfilling dreams of work and home ownership, the navvies remained in the glen above them, their lawless, sprawling encampment the antithesis of Kinlochleven's bold attempt at industrial and civic harmony. For a time it was an uncomfortable coexistence: the model company town trying hard to break free from its frontier associations yet regularly beset by episodes of drinking and brawling. Of course the navvies would eventually move on, shifting somewhere else when the work finally ended, leaving behind the village they had toiled hard to build but would never be a part of.

There was a bothy marked on the map that I could head to, a further eight kilometres to the north-east of the Blackwater

Reservoir. I crossed to the far side of the dam, watching its steep wall change colour as I walked: black from a distance, coppery brown up close and, finally, silvery white as the late-afternoon light angled off its damp concrete. Before my trip, I had come across several photos of the dam, none matching the benign conditions during my visit. Most of the images had been taken in winter. They showed easterly gales, white horses spuming on the surface of the reservoir, water spilling over the dam's upper ramparts, crashing like waves on a breakwater. I had thought of the labourers enduring these situations, caught between the duty of their work and their instincts to survive.

The distance beyond the reservoir was a hellish untrodden terrain of dark, featureless bog. Within a quarter of an hour my legs were soaked to the knees, and I'd moved less than half a kilometre. Each movement was energy-sapping, accompanied by the constant *plurpp* and *sluuthup* of plunge and release. It felt like walking at altitude, the continual hauling of legs and hoisting of feet a similar meagre return for considerable effort.

Eventually the ground began to rise and harden, and I found a path that ran above a deeply cleaved burn, all but invisible in the landscape. I then noticed something strange on the opposite bank. Rising from the moor, several hundred metres away, was a large cross, stone and starkly vertical. I scrambled down the steep incline and waded through the cold water to reach it.

It was a memorial. A long granite column topped with three splayed arms, forming the shape of an iron cross. It had been built on a pile of boulders neatly stacked into a small mound with grass sprouting from the gaps. There was an inscription at the sides of the base, chiselled in a ramshackle font of broken sentences. The words were hard to read, jumbled on top of each other, but a name was visible – the Reverend Alexander Mackonochie – a Church of England priest who perished at the spot in the winter of 1887.

On a cold but bright December morning, the Reverend Mackonochie had set out from the residence of his friend, the Bishop of Argyll, accompanied by the bishop's two dogs, a terrier and a deer hound. After issuing a goodbye to the housekeeper, who insisted he at least take some food and a walking stick, he tracked a course up the River Leven towards higher ground. It was an area he knew well, a restorative place he would often escape to, far away from the persecutions he had faced in his ministry.

Mackonochie, much loved by his congregation, had enraged the Anglican hierarchy with his use of Anglo-Catholic rituals. After many years of prosecution and lawsuits his mental health had declined and he had been forced to resign from his duties at his parish of St Alban's, London.

It is likely that while on his walk his memory failed him. Friends had expressed concerns for several years about his increasing episodes of forgetfulness and confusion. Higher in the glen, where the land levels out and where the dam is now situated, it is possible that he became disorientated, confused by the landscape and the descending darkness, heading further east instead of west to return home. A search party was raised later that evening, but to no avail. Groups of men and dogs continued looking over the following two days, but the weather deteriorated. Ill-equipped, lost and unable to find shelter, the Reverend Mackonochie perished in the conditions. His body was eventually found on the third day, located in a small hollow, the snow within it compacted from his footsteps as he had tried to keep warm. The dogs had stayed with the minister, unwilling to leave his side, and were found, cold but alive, sitting next to the body.

The rain clouds finally reached me, and I hurried the final kilometres to the bothy, keen to find warmth and shelter. Loch Chiarain Bothy was set at the head of a small loch, its gable end tall and windowless. A smell hit me before I reached the building, the stench of animal decay, sharp enough to knock my head back with the first whiff. I held a hand across my face as I passed by two sheep carcases lying half covered with sods of turf in front of the building's entrance.

I moved quickly inside, through double doors that opened onto a hallway with two large rooms either side. I called out a hello. Silence. The place was empty and full of dusty light. It was the biggest bothy I had ever been in. Up a bare wooden staircase that squeaked under my weight, there was a landing and two further rooms. Through the floorboards I could see glimpses of the rooms below, rectangles of light from each of the windows slanting through the dark spaces.

I chose one of the upper rooms, unpacking my sleeping bag onto a floor splattered with candle wax. The rank smell sifted in from outside, carried on draughts that heaved through the building. I went to collect water, moving some distance from the bothy but coming across more death, this time a deer hind, its pelt pockmarked with deep, irregular holes. Then the wing frame of a bird, thin bones aligned in beautiful parallels.

The place had got me spooked long before I heard the first bang. Back in the bothy in the half-world moments before sleep, rodents scuttled across the floorboards at ear level, the midsummer night fading to blue dusk outside. Slam! The noise billowed up from below, ringing loudly in the quiet seconds that followed. I staggered down the stairs expecting to greet a fellow resident, or to secure the door that I had left open. Neither. The door was held fast and the rooms below empty.

Lying back in my sleeping bag, I did my best to rationalise it. I was used to the sailing-ship creaks and groans of these old places. But then, minutes later: slam! The same noise again. This time I didn't care what the explanation was. Frantically, I pushed everything I had with me back into my rucksack, picked up my boots and left the building without looking back.

That night I slept out on the moor, nestled in the heather, away from the gut-pulling stench of decay and away from the tall, looming bothy. The next day I would make the long journey back to the reservoir, past the monument – that strange testament to person and place. I would cross at the dam, but not revisit the graveyard, its abject isolation enough for me once. I would climb back up towards the Devil's Staircase, listening to skylarks and counting the hours of my own small migration. But before then, I would lie under a pale, starless sky, my face exposed to the night's drizzle.

The Memory Path

Jock's Road, Braemar to Glen Clova

Make a journey south from the Highland village of Braemar, following the riverbanks of the Clunie Water and then the Callater Burn, and you will enter a place of wildness and ancient passage. For a path exists here, more remembered than real in places, incomplete and at times imagined.

Continue on, climbing higher, tracking burns that vanish in the gradient as the trail appears and disappears: there and not there, glimpsed briefly in lines of inky footprints only to be brushed clean in the yellow deer-grass tundra of the Mounth plateau. Maps can be deceiving here. The confident black dashes, stretching out across the grid squares indicating a track, feel fictional at times. Instead, a route must be felt through the landscape, tracing the idea, rather than the certainty of a path.

It was a crossing I had wanted to make for years. A traverse through empty glens and across the tops of desolate, whale-back mountains. I would walk it north to south, the way it had once been travelled, on land pressed down by centuries of human and animal footfall. I wanted to feel something of the weight of that passage, to know the ritual and the routine of the drovers that had once used it. I was fascinated by the ghostliness of it all. The years of movement in the landscape, a practice passed through tens of generations, now so completely forgotten.

The route had other lures as well. Despite its remoteness, it seemed somewhere of recurring notoriety and significance, a place of overlapping histories. For it was here that a bitter class battle had begun. A confrontation which sparked a dispute so acrimonious it would come to be a defining moment in the rights of access to Scotland's wild places. Tragedy and loss were also present on this route. In 1959, near Jock's Road, the highest part of the pass, five mountaineers lost their lives during a ferocious midwinter blizzard – at the time, Britain's worst mountaineering disaster. Some years later, a refuge was built, erected in the memory of the walkers and with the intention of saving the lives of other future travellers to this place. I had heard the shelter was both primitive and tiny in its construction, built into the landscape like a Hobbit hole. It was called Davy's Bourach and, by my reckoning, was one of the highest usable shelters in Scotland. It was, I had decided, somewhere I had to stay on my journey.

The morning was warm, and the perfumed scent of heather pulsed across the open moor. It had not rained in weeks, and small dust clouds billowed up with every footstep as I tramped the long vehicle track into Glen Callater. On either side, the grass had a parched appearance, yellow and brittle, rasping in the breeze. Above me a buzzard circled, eyeing my slow progress through the valley, the outline of its wing tips quivering in the hazy light.

I had been walking for just over three hours and met only one other person. He had appeared at a sharp bend in the track, and the unexpected sight of someone else took us both by surprise. He had the loose, trodden-in look of the long-distance walker, stooped slightly under the weight of his pack, his skin deeply tanned. He told me in a heavy German accent and a quiet voice that he had walked from Banchory the day before. It was, he nonchalantly mentioned, part of a journey he was making to reach the Atlantic coast. He had no timetable, he said, just the aim of covering a decent distance each day. I envied his lack of schedule, knowing I had only two spare days to make my crossing.

Loch Callater eventually came into view: a scimitar shape of glinting water, metallic and sharp between folds in the moorland. Ahead of it, I could see a set of buildings, crowded close together with a narrow passageway running in between. A fence had been erected around the plot with saplings planted at the boundaries and a tall, white flagpole placed at the entrance. These were the Callater Stables, part of the vast Invercauld Estate and once the last refuge for drovers before they made the crossing of the plateau.

As I approached the buildings, two mountain hares crossed the path ahead of me, one chasing the other in a wide arcing run, all ears and spindly legs. The pursuing hare stopped and sat bolt upright, suddenly aware but unconcerned by my presence, only bothering to move when I came within a dozen paces.

The largest of the buildings was boarded up. Pea-green panels, the colour of vintage double-decker buses, were fixed tight against the windows. I tried the heavy-set door, but it was bolted fast with a padlock. A horseshoe was nailed at the top, encircled by the twelve-point antlers of a decorative stag's head, the name 'Loch Callater Lodge' written in the middle. To the left of the entrance was an old iron boot scraper, rusted to a deep orange. Adjacent to the lodge were two further buildings. One, a large wooden shack with a corrugated roof, the other, a small stone bothy, the door left ajar. I checked to see if anyone was inside.

The place was empty but held the feeling of recent occupation. In the larger room, the floor had been freshly swept, and bright plastic chairs had been stacked neatly against the wall. In the thick window recess, someone had left a camping kettle, a half-finished bottle of beer and the shells of tea lights whose wax had melted away.

On the table I found pencils, water bottles, hand sanitiser, washing-up liquid and more diminished candles. I also spotted the bothy's visitor book. The place, it seemed, was well used – most people staying here as part of a larger journey, resting and moving on as the drovers had once done. I read descriptions of long-distance hikes, cycle tours, and arduous solo walks, all undertaken by more nationalities than I could count, each retelling their journey to the bothy and their experiences while staying there.

Some entries covered entire pages of the A4 pad, supplemented with drawings and doodles. One gushing note had written thank you several times within interlocking hearts. Another talked of the multitude of mountain hares they had seen and included a drawing of one in a heraldic pose, sitting sejant on its haunches. I wondered if it was the same animal that I had seen, a guardian of the place perhaps. Other comments were less sentimental, one simply reported, 'walk, wind, bothy, lunch, left'. I was, however, unable to work out the many references to the bothy's excellent internet connection, a running joke throughout the book that was lost on me, until I noticed a pilfered sign above the desk, promising 'Free Wi-Fi here!'

Something was missing in the bothy, though, and I struggled for a few moments to put my finger on it. The place was functional and well equipped, with two bunk beds in the smaller room, as well as the larger communal space. But there was an incompleteness I couldn't quite grasp. I sank into an old leather armchair, the springs popping and chiming under my weight, and pondered what was lacking. Suddenly, it dawned on me. Of course – there was no fireplace. I couldn't remember ever visiting a bothy without one, and its absence deprived the place of a certain homeliness, let alone warmth. In winter, I imagined the building to be desperately cold, the thick stone walls turning the inside into an ice house for months on end.

Back outside, I sat on a low bench in the sunshine, my back resting against the whitewashed bothy. It felt strange to be among the cluster of buildings, so deep in such a wild spot. Stranger still to be there alone. The place seemed to carry an echo: sounds of arrival and welcome, of movement and rest – a residual hum of past human activity that was all the more audible in the solitude.

John Elphinstone's *New & correct map of North Britain*, published in 1745, is one of the most detailed maps for the period. At a time before the full devastation of the Clearances had taken hold, it shows Scotland as a densely populated country, chock-full of place names. Townships are abundant, marked throughout the Highlands and the Lowlands, bristling along coastlines and packed so deeply inland that the writing bends and contorts to fit the available space.

Overlaid across the map are also topographical features. Mountains appear in relief, small pimples scattered across the chart, their eastern profiles illustrated with delicate cross-hatching of shade. Rivers and lochs are profuse, a complex fretwork running either side of the watershed, spread out like the veins in a leaf or the creases on a palm.

It is only when you look closely that you can see something else. A few faint lines, depicting roads that connect the cities of Edinburgh and Glasgow and a handful of garrison towns: Fort William, Ruthven, Inverness, Dalwhinnie. The visual linkages are subtle, impressed lightly on the paper, as if they held less subconscious relevance in the mind of the cartographer.

In this sense, the map reveals something of its true priorities. It describes placement rather than passage. A record of a time when journeys were still determined by the natural features of the land, resolved along lines of least geographical resistance – through glens, around mountains and alongside rivers. Roads – apart from the military highways laid down to suppress the Jacobite rebellions – were primitive, often no more than causeways of stone or gravel, and in Elphinstone's opinion, certainly not worth the ink to portray on a map.

Riding eastwards over a hill into Glen Tilt in 1769, the Welsh naturalist Thomas Pennant described the difficulty of travel on such undeveloped highways. 'The road is the most dangerous and the most horrible I have ever travelled; a narrow path, so rugged that our horses often were obliged to cross their legs, in order to pick a secure place for their feet; while, at a considerable and precipitous depth beneath, roared a black torrent, rolling through a bed of rock.'

Yet by the latter part of the eighteenth century, something amounting to a mass migration was taking place every year across Scotland. Huge numbers of cattle and sheep, and the people who drove them, streamed inwards from every corner of the country. From late spring to early autumn, hundreds of thousands of animals were moved from their breeding grounds to the cattle markets of central Scotland and then, more often than not, south of the border to English towns and cities. The journeys would have been time-consuming and arduous, negotiating everything from sea-crossings and mountain passes to moorland bogs and river rapids. Little more than ten miles (sixteen kilometres) would be averaged on a good day, with many animals not surviving the experience. The success of such an enterprise rested firmly on two factors: the skill and resolve of the drovers, and the existence of a network of drove routes to enable a relatively efficient and safe passage.

Many of Scotland's smaller drove routes had already been in existence for millennia, created from the seasonal movements in the grazing of cattle. This ancient transhumance had occurred from the earliest rearing of livestock in Scotland, resulting from the need to alternate pastures – from higher ground in summer to the more temperate, lower valleys in winter. Time-worn passageways evolved, small regional byways, 'green lanes', 'ox roads' and 'drift ways' that described continual, repeated acts of travel so ancient it would be impossible to date their original use.

Understanding the timeframes over which these routes were created stretches our sense of human and landscape history. 'It is impossible to state how long it would take the feet of herds to wear down the hard oolite stone by even one inch', wrote the historian K.J. Bonser, describing a set of cattle tracks worn several feet into the ground. 'To measure this age-long traffic by centuries is inadequate; the age must be measured by thousands of years.'

By the Middle Ages, these short, localised tracks were becoming increasingly connected with neighbouring districts, creating longer-distanced paths which, when linked together, formed capillaries of cross-country routes throughout Scotland. Over time, from this network of informal pathways, a distillation of 'known' routes established themselves: those tracks that through generations of collective, personal experience were acknowledged to provide the safest and fastest transit. Yes, a particular mountain pass may appear to lead most directly to an adjacent glen, but local familiarity with its weather-prone slopes may dictate a longer circumvention to be more reliable.

In this way, orientation and route-finding had become subjected to a wayfinding equivalent of natural selection. And from around the fourteenth century onwards, this knowledge enabled the practice of droving to flourish commercially in Scotland for hundreds of years, despite the lack of any formalised road networks.

That's not to say the journeys were easy. Over such long and elevated distances, storms would at some point be a certainty, as would the prospect of armed cattle theft. Let alone the logistical difficulties that would be faced. Moving hundreds of head of cattle over vast areas of inhospitable terrain, all the while ensuring they remained well fed, uninjured and ready for sale at the end of it, was a massive undertaking. It called for individuals with not only a specific set of technical skills, but also a particularly hardy disposition. 'The characteristics and qualities required of a successful drover were many,' wrote A.R.B. Haldane in his book *The Drove Roads of Scotland*. 'Knowledge of the country had to be extensive and intimate, while endurance and ability to face great hardships were essential.'

According to many contemporary commentators in the eighteenth century, the Scottish (and in particular, the Highland) population was already predisposed to this kind of occupation. 'He has felt from his *early* youth,' wrote Sir John Sinclair, compiler of the *First Statistical Account of Scotland*, 'all the privations to which he can be exposed in almost any circumstance of war. He has been accustomed to scanty fare, to rude and often wet clothing, to cold damp houses, to sleep often in the open air or in the most uncomfortable beds, to cross dangerous rivers, to march a number of miles without stopping and with but little nourishment, and to be perpetually exposed to the attacks of a stormy atmosphere. A warrior thus trained suffers no inconvenience from what others would consider the greatest possible hardships.'

Sinclair's veneration of drovers as a kind of 'warrior-class' profession is perhaps unsurprising. Driving cattle, indeed, resembled a military campaign, and the attributes required to do so were very close to soldierly. 'The Highlanders in particular,' commented Sir Walter Scott, 'are masters of this difficult trade of driving, which seems to suit them as well as the trade of war. It affords exercise for all their habits of patient endurance and active exertion. They are required to know perfectly the drove roads which lie over the wildest tracts of country.'

By making such a worthy comparison, both Sinclair and Scott legitimise and begin to romanticise qualities associated with Highlanders that only decades before would have been vilified and outlawed. In droving we can see a socially and politically acceptable version of activities that for centuries had (on the whole, unsuccessfully) been prohibited. Hereditary experience of clan warfare, armed insurrection and, most notably, cattle reiving (stealing) then came to be valued as having a practical, commercial application in the skills required to drive cattle. 'To a Highlander of the eighteenth century, divided at most by one generation from such a way of life,' wrote Haldane, 'and possessing beyond a long lineage of cattle-reiving ancestors, it was but a short step to a more legitimate and only slightly less adventurous form of cattle driving.'

This tacit acknowledgement of the expertise of the drovers reflected a growing economic imperative. With the industrial revolution gaining momentum and with increasing numbers of workers heading to the cities, the need to sustain Britain's urban populations became paramount. Droving would reach its zenith during this period. At a time before road and rail networks had opened up the country, the drove routes became essential to the prosperity and progress of the expanding British Empire. For a while at least, the workshop of the world depended on these age-old pathways, and on the ancient knowledge of their usage.

It had been a mistake to rest as long as I had. The long walk in the unexpected heat had left me tired. My legs ached and my back felt stiff. For a moment, I wrestled with the temptation to stay longer at the stables. The weather, though, was beginning to change. The sky remained cloudless, but a breeze had picked up from the south, pushing warm air against my face and compelling me to move.

I tracked a path south-east and climbed a small knoll that overlooked the buildings. At the top, I found a large cairn that had been built only a few years earlier: big lumps of granite, still free of lichen and moss, cemented together on a square concrete base. At the front was a rectangular tablet, salmon-pink and shiny, commemorating the Queen's diamond jubilee. It seemed an unusual tribute to have erected there. Almost every memorial I had come across in the mountains had been to mark a life tragically lost. I thought back to Mackonochie's stone column near the Blackwater Reservoir, but also to the plaque to the air-crew lost in the Cairngorms that I had discovered high on the empty plateau of Beinn a' Bhuird, and the huge iron cross on the summit of Ben Ledi, in memory of a member of a local mountain rescue team killed on duty. Each memorial had, despite – or because of – its placement in a wild landscape, resonated a powerful sense of place and poignancy.

I crouched next to the cairn, leaning back against its weight, and bit into an apple. From up high, the Stables below had an even more desolate feel. I imagined the place deserted, like an abandoned colonial outpost, crumbling away at the edge of the Empire. The empty flagpole, the broken boundary fence and the cairn next to me evoked a Victorian sense of hubris and self-belief. A mix of the missionary and the misguided, the attempt at imposing civilisation on an indifferent wilderness.

My route skirted the edge of the loch, following a path made narrow by the encroaching heather. I had to walk with one foot directly in front of the other, as if balancing on a tightrope. Where the track met the water, I stepped across miniature beaches of pink shingle, my boots sinking softly with each step. At a larger cove, I came across a prism-shaped boulder, rising several feet from the ground at an area marked on my map with calligraphic font as the 'Priest's Well'.

As the story goes, the prominent rock, known as the Priest's Stone, is so called because of a local clergyman. One particularly bitter winter, the Braemar area experienced temperatures so cold and so prolonged that the ground remained frozen long into spring. With no sign of a thaw in sight, and with food and water becoming scarce, the local people called on a holy man, named Phàdruig, to help. The priest led his congregation to the stone, which marked the site of a sacred well. The spring was frozen, but as Phàdruig began praying, the well ran clear, and as the ice melted here, so too the rest of the district became freed from the winter's grip. The priest's miraculous intervention became an enduring local legend, and his name is commemorated and remembered elsewhere in the glen. There is Creag Phàdruig, the rounded hill directly north of Callater Lodge that holds the high tarn of Loch Phàdruig, and to the east of the loch, there is the munro Carn an t-Sagairt Mor, or the priest's hill.

The wind had gathered strength as I reached the far end of the loch. White horses broke across the open water, and small waves carrying a foamy flotsam of grass and heather stems lapped against the shore. I eyed the clouds that had begun to build over the hills in the south, trying to gauge their size and form.

Entering the upper reaches of Glen Callater felt significant, as if I were crossing a marker point. The landscape had taken on a rougher, more elemental look. The surrounding slopes were now closer, and the wide vistas of the early part of my walk had been replaced by narrow sightlines: sharp, overlapping angles and the falling diagonals of converging ridges. Colours had also changed. The yellow-green of moor grass had ceded to the purple-browns of old heather and the deep black of peat hags. At one point, I passed a wide cross-section of dark earth, sliced cleanly from the ground above it. Inside the trench, rising from crumbling soil, was an ancient tree stump, its roots marble-white and splayed around it like tentacles.

The path – now, for the first time, referred to as Jock's Road on the map – came and went, withering in softer ground, and then reappearing confidently as it tracked the side of the Allt an Loch stream. I became used to its temporary presence, but by the time I had reached the headwall of Coire Breac I had lost sight of it altogether. It was then that I saw the post, fifty metres ahead of me, made of metal and rising on a perfect vertical.

I did a double-take, not expecting such an abrupt intrusion on the wildness of the place. It should have been a signpost, but its pointer was missing. My guess was that it had been placed there to indicate both the direction of travel north towards Braemar and the route to Glen Doll in the south. It seemed a somewhat unnecessary navigational aid, considering the only routes were either through the confines of the glen or up onto the plateau. That, though, was not really the point. The signpost was a statement, a declaration of a bitterly-fought-for right whose story began on the Jock's Road path.

'I did not ask anyone's leave, because I did not think I required it,' replied the Reverend Grant Duncan when asked during court proceedings in 1887 if he had obtained the landowner's permission before attempting to walk Jock's Road. 'I have understood all my life', he continued adamantly, 'that that was a public right of way.' The Reverend Duncan's response was forthright and unyielding, delivered with all the certainty of a man of faith. Whether or not he knew it at the time, it also captured the essence of what was to be a crucial legal challenge regarding the public's right of access to Scotland's, and later Britain's, wild places.

For centuries, an informal right of access had existed across Scotland, allowing an unquestioned freedom of movement across practically every corner of the country. But by the beginning of the nineteenth century, this accepted notion of unfettered access to the land was becoming increasingly threatened by the commercial self-interest of the land-owning elite.

By the early nineteenth century land use in Scotland had entered a period of shameful moral upheaval. The Clearances, although undefined by a specific starting date, were well underway, with many landowners across Scotland looking for a more profitable return on their often vast, but largely unproductive, land. The rearing of sheep, displacing the far more bothersome human tenants, was the obvious solution and became commonplace across the country. However, the idea of another land enterprise, which would have similarly punitive effects on local populations, was also taking hold.

Records of large 'sporting estates' as a means of commercial venture can be traced back as far as the late eighteenth century. The premise was simple and predicated on taking advantage of an Act of 1621 which enshrined a legal relationship between landownership and the right to hunt game. While the Act determined the position of '*res nullius*' (that no one individual could claim ownership of game, be it deer, grouse, ptarmigan or hare), the right to hunt quarry rested solely with the owner of the land on which the animal was found. A right, landowners realised, that could be leased for a handsome fee.

The money-making idea quickly took hold. Land-owning magnates across Scotland hurriedly seized the opportunity to convert their hills, moors, rivers and forests into sporting playgrounds for the new business elite of the Industrial Revolution. By 1812, Highland lairds such as the Duke of Gordon had begun advertising hunting as a (paid-for) sporting activity on their estates. The number of large 'deer forests', in which animal numbers were actively maintained for game purposes duly rose from around six in 1811 to twenty-eight in 1839, then to fifty-four twenty years later – boosted in no small part by Queen Victoria's avid endorsement of the sport as well as the 'Balmoralisation' of large parts of Scotland. At the same time, large amounts of grouse moors were also being established, bringing the number of Highland sporting estates to seventy-nine by 1873.

Management of this lucrative new form of recreational capitalism was, however, ideologically at odds with the traditional notions of public access. Land, and the quarry within it, became jealously and belligerently guarded. Reports of hostile landowners and angry confrontations between ghillies and hill-goers became a regular occurrence and even began to gain wider public attention. On one of his regular field trips to the Highlands in 1847, John Balfour, Professor of Botany at the University of Edinburgh, was famously stopped from venturing into Glen Tilt. The professor was met by the Duke of Atholl and a retinue of his men, who barred access onto the land. An angry exchange ensued which only ended when the professor and his students vaulted a dyke and ran off into the glen.

At the same time, however, a parallel rebuttal to this exclusionary form of landownership was gathering momentum. In Edinburgh, the Association for the Protection of Public Rights of Roadway was formed to challenge access restrictions in the countryside surrounding the capital. Among its members and supporters was a considerable number of well-educated professionals, including lawyers and several members of parliament. Here was a society comprised of men and women well versed in the frameworks used for exerting meaningful political pressure. An organisation more than capable of taking on both the landed establishment and the aristocracy. Its self-ascribed remit soon expanded to defend rights of way across Scotland, crystallising in a case that would not only bankrupt the society but also determine the future of public access to Scotland's countryside.

The year was 1885, and after making his fortune as a grazier in Australia, Duncan Macpherson had returned to his native Scotland to purchase the Glen Doll Estate. Macpherson soon made clear his intentions of turning the land into a sporting estate and, in doing so, categorically keep the public out. The Association, renamed the Scottish Rights of Way and Recreation Society, however, was quick to test his resolve on the matter.

Regular forays were made onto Macpherson's estate, which were met with resistance from the landowner and his men. Macpherson responded in kind to the perceived intrusions, erecting fences with padlocked gates and planting trees across the path. The Society was undeterred and proceeded with a plan to signpost the right of way through the Mounth and Glen Doll until they were intercepted by the estate keepers. This final piece of direct action brought things to a head for both parties, and a court case was soon to follow. The Society issued a summons against Macpherson, claiming that 'there exists a public road or right of way' across Jock's Road, and they 'and the public generally are entitled to the free and lawful use and enjoyment thereof'.

It was a bold and uncompromising move, but one founded on robust legal logic. If they could demonstrate that the route was a passage of such long-standing use, then – the Society reasoned – no landowner could revoke such an ancient right of way. They did so with diligent efficiency, mustering dozens of testimonies from people who had once regularly walked the path, including the Reverend Duncan. The Court of Session papers in Scotland's National Records Office list each of these witness statements. Pages of transcripts, written in the elegant handwriting of the court recorder, detail accounts from crofters, farmers, drovers and shepherds, and even a teacher and a travelling tailor.

To read the statements is to understand something of the imaginative resonance of wild places, for the words are humbling in their intimacy of recollection, despite the passing of time. In many cases, decades had gone by, but still the ability to reminisce about the landscape remained undiminished in those questioned: the sweep of a particular mountainside, the way a burn described its course through the hills or the point at which a waterfall tumbled. For these men, the path had endured, mapped in their minds just as clearly as it was physically defined in the land.

It was these unadorned but compelling testimonies that eventually swung the dispute in the Society's favour. The action was finally settled in the House of Lords at huge cost to both the Society and Macpherson, both of whom became bankrupt. The outcome of the case, however, set a powerful precedent both legally and in the financial implications to landowners who attempted to disavow historic rights of way on their land. More importantly, the ruling also galvanised a greater public consciousness towards rights of access to Scotland's wild places. It prompted a series of legislative measures which, over the following century, were to lay the foundations for our modern-day exploratory freedoms.

There was only one way out of the corrie and up onto the plateau: a steep climb between the outcrops of two peaks, Creag Leachdach and Tolmount. I gained height quickly as I walked, stopping regularly to plant my palms on my knees and draw in breath. The clouds I had seen building an hour before had now reached the tops and were descending as I was ascending, spilling into the corrie and sinking like dry ice into deeply cleaved burns. I edged alongside a series of cataracts as my map suggested, looking for the shallowest incline. The waters ran fast at first, kinking and bending down the hillside. But eventually as the higher, flatter ground was reached, they thinned away to nothing, and I was left directionless on a wide expanse of blank mountain moor, buffeted by cloud-filled squalls.

For the drovers, lacking reliable maps and navigational devices, route-finding in these conditions would have been difficult, bordering on treacherous. The Jock's Road route was particularly dangerous, combining, as plateaus do, the hostile characteristics of height and featurelessness. 'This is one of the most serious walks described in this book,' states the description for the route in the 2004 edition of the *Scottish Hill Tracks* guide. 'It crosses a high, exposed and featureless plateau which in winter is frequently swept by storms. At that time of the year the path over the plateau is likely to be covered by snow and completely invisible for several months.'

Once on the high ground of the Mounth, a reprioritising of senses takes place. The power of sight is diminished: depth of field impaired in the vast areas of uniformity, or lost completely in the white blindness of cloud cover. Meanwhile, hearing and touch are amplified: wind howls at greater decibels, and the cold becomes more penetrating.

Finding one's own way in this environment is hard enough. There exists the ever-present prospect of becoming lost in kilometres of bewildering similarity, or worse still, finding one of the plateau's many edges – the cliffs and crags that mark an abrupt and irrefutable exit from height. Even greater attention would have been required from the drovers. Marshalling a herd of anywhere between forty and four hundred head of cattle must have called upon immense reserves of patience and sound judgement.

There were certainly easier routes than Jock's Road. The alternative path south from Braemar, for instance, up Glen Clunie's long, grassy valley and then over the

Cairnwell Pass to Glen Shee, offered a lower, more gradual passageway through the Mounth Range. This route, however, would have been busy, and after the 1745 Jacobite Rebellion became part of General Caulfield's military road network. Where possible, the drovers preferred not to use such highways, partly to avoid paying tolls, but mainly to limit the damage that could be done to their cattle's feet. There was also the more pressing desire to maintain a low profile, specifically to escape the attentions of roaming bands of cattle thieves. This logic probably explains why Jock's Road remained in use for so long. Although exposed, the plateau provided ample room for cattle to feed and rest, while leaving little scope for unexpected attack.

Of course, the demands on the drovers taking this route were significant. Navigationally, after reaching the plateau from Glen Callater, they would have to correctly time a sharp change in direction – turning right and heading south so as to successfully angle into Glen Doll and not into one of the neighbouring glens. Such was the importance of this task on routes like Jock's Road that larger droves would include a 'topsman', a specialist pathfinder who would break trail some distance ahead of the herd, guiding the safest passage.

Hardships would also have to be endured. Drovers would sleep in the open air with their cattle, fully exposed to the mountain weather and wrapped only in blankets or (more commonly) just the plaids they wore during the day. A night watch would also need to be kept, ensuring no animals strayed or were stolen. The prevalence of armed reivers in many parts of the country, including the Mounth, was so common that drovers carried a variety of weaponry to defend themselves, from broadswords and poniards to pistols and muskets. The threat was indeed so endemic that drovers were exempt from the Disarming Acts of 1716 and 1748 and even allowed to bear arms during the period of the 1745 Rising.

Provisions on the more remote stretches of a drove were limited to what could be carried, compounding the privations experienced. Walter Scott describes the typical drover's diet as 'a few handfuls of oatmeal and two or three onions, renewed from time to time, and a ram's horn filled with whisky'. In areas such as the Mounth, with scant wild food or fuel available, there would be little opportunity to supplement this meagre fare. Instead, if required, drovers would bleed their animals, mixing this with their oatmeal to make black pudding.

I walked through the mist, trying to find landmarks from which to fix a compass bearing. The air around me was a bright pearly white. The clouds had rubbed the nearest peaks from view and reduced visibility to a couple of hundred metres in every direction. I gave up on finding any sign of the path and instead headed east into the centre of the plateau, counting my paces and trying to estimate the distance before I needed to turn south. Through more luck than judgement, I eventually found signs of previous passage: at first a thin channel, worn between the tussocks of deer grass, then a set of smudged footprints. I was soon on a small, muddy track, certain I had rejoined Jock's Road. The landscape also began to fit the descriptions on the map, rising almost imperceptibly to the highest part of the route, the broad summit of Crow Craigies. I felt reassured in my direction, but the plateau's weather was gathering energy. The wind had intensified and the sticky, saturating drizzle had morphed into fast-flying rain bullets that thudded diagonally against my jacket, drumming on my hood.

In the two kilometres it took to reach the shelter I became thoroughly soaked. Eventually I saw the refuge – or rather, saw where it was located. Two red marker poles, three feet high, stuck out from what appeared to be a mound of heather next to the path. I rounded the embankment and saw a small door, post-box red, tucked into a wall of boulders. Above it was a corrugated iron roof, overgrown with heather and grass so that only its frilled, rusting edges were visible. Attempts had been made to secure the metal sheeting against the storm winds that regularly funnelled up through Glen Doll. Rocks had been placed in the centre of the roof to provide ballast, and wire cabling lashed it down at the sides.

An elaborate sign, embossed with a hillwalking club emblem, was riveted to the door, requesting for the refuge to be left shut and to 'leave no litter'. I slid the latch across. The heavy door pivoted noiselessly on its hinges, revealing a space so dark I needed my headtorch to see inside. I unstrapped my rucksack and stooped low to fit my head and shoulders through the entrance. The place was surprisingly large. Too low to stand up in, but big enough to sleep a small group of walkers.

At head height, several wooden cross-beams spanned the width of the shelter supporting the roof. As I scanned my torch across them, I could see the names of previous occupants written and carved into the timber. Apparently, the refuge had been frequented by, among others, the 'Glesgovian Bothy-Billies' in 2006, the 'Misty Mountain Hopper' in 2009, and in the summer of 2010, Ryan (aged nine).

I scouted the floor of the shelter for the best place to make camp. The ground was damp, and in the places where rainwater had dripped in I could see small runnels of silvery mud. It was, though, remarkably dry inside considering the high-velocity rain strafing the moorland outside. Without much thought, I busied myself with a familiar routine. I hung my sodden clothes from nails in the beams, heaped on the extra layers from my rucksack and began heating pasta on my stove. I lit candles, placing them in small recesses in the wall, and watched their flames being bullied by drafts, flickering and bouncing light around the shelter.

I felt grateful for the refuge's temporary sanctuary, especially so high on the plateau. At almost 2,800 feet, it was the highest usable shelter I had found anywhere in Scotland's mountains. Constructed in 1966 by a renowned local hill man, Davy Glen, it affectionately became known as Davy's Bourach, referencing both its builder and its rather ramshackle appearance – 'bourach', a Scots word meaning a muddle or a mess, but also a small home. It was built near the site of another, far older shelter, known as Jock's Hut. The eponymous Jock was reputed to be a local shepherd, John Winter, who gained notoriety by taking sides in a land-owning dispute between Lord Invercauld and Lord Aberdeen. As the story goes, the shepherd encountered Lord Aberdeen one day on the drove road and refused him passage across the Mounth. Perhaps later realising the consequences of his action, the shepherd then went into hiding from the laird, living out on the high ground that now bears his name.

By the 1960s Jock's Hut had all but disappeared, and its ruins were used by Davy Glen to build his shelter. In a feat of considerable individual effort, he carried up the other necessary building materials himself, shouldering spars and roofing panels along the steep path from Glen Doll. The undertaking was not only testament to Davy's physical resolve but also a labour of love, an endeavour spurred on by a tragic event that had taken place near Jock's Road seven years earlier.

On New Year's Day 1959, five members of the Universal Hiking Club, a Roman Catholic mountaineering group, attended the morning mass at St Andrew's Church in Braemar. They had been exploring the surrounding area as part of a large club meet, and their intention that day was to meet up with fellow club members at the youth hostel in Glen Doll. They planned to make the journey on foot, covering a distance of more than twenty kilometres by following Jock's Road across the Mounth.

By the time the men had attended the church service and left the village, it was already late morning. Heavy rain and driving sleet soaked their clothing soon after they began walking, and a biting wind blew in from the north. They persevered, undaunted by the miserable conditions, and were seen by a sheep farmer around noon making their way towards Glen Callater, still with the majority of their walk to complete, but with little more than four hours of daylight left.

Understanding exactly what happened later that day is hard to say with any degree of certainty. It's known that the five friends continued on the long walk through Glen Callater to eventually reach the steep confines of Coire Breac. For some reason, instead of following their intended route up the steep headwall of the corrie tracking the course of Jock's Road, they took an unplanned deviation. Footprints later found by one of the search parties looking for the men indicated they had climbed up the south side of the Tolmount peak.

From here, conditions would have become atrocious. The rain and wind they encountered at lower levels had turned into a blizzard of extraordinary proportions. Reports from elsewhere that day speak of one of the worst winter storms in years. On the nearby summit of Meall Odhar, skiers described winds of up to 100 kilometres an hour. Roads in the area quickly became impassable from snowfall, and temperatures dropped well below freezing.

For the men, breaking trail through heavy snow, battling against Arctic winds and with darkness descending, the situation was dire. Within two kilometres of the start of the descent from Tolmount, the first of the walkers, Frank Daly, succumbed to the cold and died. His friends had little choice other than to continue towards lower ground, but were themselves unable to go much further. Within 400 metres, another member of the group, Robert McFaul, was also overcome, and shortly afterwards the club's hiking convener, Joseph Devlin, fell through the snow, becoming trapped in the White Water Burn.

The final two men, the club's chairman, Harry Duffin, and seventeen-year-old James Boyle, managed to walk another couple of kilometres following the line of the stream, only to reach a narrow gully and waterfall that dropped precipitously into the upper reaches of Glen Doll. The Jock's Road path tracks a course several hundred metres north of this exposed feature, allowing for safe access into the glen. But in the disorientating conditions, negotiating such a difficult exit became impossible and both Duffin and Boyle died in the attempt.

By the afternoon of the following day, with the party failing to arrive at the youth hostel, there were serious concerns as to their whereabouts. For more than three days a major search and rescue mission was mounted. Over a hundred people were involved, including gamekeepers, shepherds, skiers and climbers, as well as several mountain rescue teams and the police. Conditions, however, remained appalling. 'The weather was so severe, with the wind whipping the snow into a turbulent blizzard,' wrote Ian Thomson in his definitive account of the tragedy in his book, *The Black Cloud*, that 'the searchers could make progress at times only by bending low and turning their sides or backs into the storm to gain some respite from its fury'.

It quickly became apparent that the chances of the men surviving were slim. On the second day of the search, a party on the Glen Doll side of the pass found the body of James Boyle in a gully near the White Water waterfall. The rescuers continued in their efforts for a further two days until it was concluded that all hope of finding the others alive was lost. It would not be until a thawing in the snow cover that they were found. Davy Glen, who had participated tirelessly in the initial search, and had continued to look for the men during January and February, organised another search party at the start of March as the weather became milder. It was then that the body of Harry Duffin was found, followed in the weeks after by the discovery of Robert McFaul and Joseph Devlin. With remarkable compassion and determination to find the men, Davy Glen was involved on each occasion.

It was not until April that the final man was found, when members of the Carn Dearg Mountaineering club came across the body of Frank Daly, lying where his friends had left him on the south side of Crow Craigies. It drew to a close one of the most tragic episodes in British mountaineering history. A heartbreaking event that is now commemorated by a small metal plaque, fixed discreetly to a large boulder high on the Jock's Road path. Another memorial in a wild and empty place.

I drifted off to sleep listening to the horror-movie chuckles of ptarmigan as they roosted in the twilight moorland outside. Moments later (although probably hours had passed in the compressed timelessness of the shelter), I was woken by rainwater on my face. Drips falling in an even staccato rhythm. Without leaving my sleeping bag, I shuffled across the ground to find a drier spot. Soon afterwards, the same thing happened again, and I repeated my bug-like slither to a different place. I worked my way round most of the shelter that night, finding out that I had an uncanny ability to position myself under each new leak that came through the roof. Sleep was minimal, but in the bouts of wakefulness I felt secure in the darkness, content in my containment, and understanding something of the poet Norman MacCaig's assertion, that 'such comfortless places comfort me'.

Outside, dawn was exactly the same as dusk – as if only minutes, instead of a night, had elapsed. I was still within the clouds: daybreak muted to a grey, papery light, the rain heavy and directionless. Beyond the shelter I crested a small ridge, the path dropping through gaps in the rock by the

White Water waterfall. I tried to imagine the drovers here. The route was so narrow that their animals would have moved in single file, a long procession patiently cajoled, the dark edge of forestry finally in sight.

Wild Island

Inishail, Loch Awe, Argyll and Bute and *Eilean Fhianain, Loch Shiel, Lochaber, West Highlands*

Of all of Scotland's wild places, its wild islands enthral me the most. Perhaps it's some relic from childhood. The consequence of reading so many children's books that imbued islands with a sense of mystery and adventure. They were the realms of exploit and exploration, wondrous territories described in a tantalising prose of danger and discovery. Places such as Robert Louis Stevenson's *Treasure Island*, that could claim 'latitude and longitude, soundings, names of hills and bays and inlets, and every particular that would be needed to bring a ship to safe anchorage'. To a child's mind they also represented powerful notions of independence and freedom, distanced both physically and imaginatively from the timeframes and rules of the adult world. Most famously, in Arthur Ransome's *Swallows and Amazons*, Wild Cat Island becomes a private dominion for the children of the novel: their own unexplored island kingdom, to be mapped, named and ruled over by them, as if they are the first to ever land on its shores.

There is, then, something childlike in my fascination with wild islands, but there is also something inherently beguiling about their separateness. Their geographical detachment from the mainland and the difficulty to reach them provokes the same stubborn inquisitiveness in me as remote mountain peaks do. They are also, as I have come to discover over the years, places that have been repeatedly weighted with human importance, invested with strategic or spiritual significance through multiple generations.

I have visited dozens of Scotland's wild islands, but there are countless more I have never set foot on. Many are cast out at sea, but it is the inland varieties that really capture my imagination: wooded islets, freshwater archipelagos or tide-marked outcrops sheltered in sea lochs. None are lived upon (for to be truly wild, an island must be uninhabited) and unlike other wilderness areas, few attract much human traffic. In their containment and isolation they have become vessels of past events, water-locked archives for some of Scotland's wildest histories.

I had in mind to extend my exploration of wild islands further. For some time, I had heard of a cluster of islands at the north of Scotland's longest loch, Loch Awe, which were both mysterious and primordially wild. Forest islands, swathed in ancient pines, from whose uppermost branches raptors could be seen lifting off in flight. I planned to make the journey by canoe, sleeping out on the islands' tree-fringed shores and paddling between their narrow channels. Later, I would also make a crossing to Eilean Fhianain, a small island in the vast waters of Loch Shiel that had once been home to the Irish missionary, Saint Finnan, and where a ruined chapel and burial ground describe the religious schism that once bitterly divided Scotland.

In the middle of the canoe, brightly coloured dry bags had been stuffed beneath the thwart, wedged tightly together so they'd remain in place even if we capsized. Under the bow seat I'd crammed a rucksack full of camping equipment and enough supplies to last us a few days. At the stern was a steal fire pan and a bucket full of firewood, to be used for cooking and for warmth.

I'd been joined by my best mate, Steve, grabbing the chance for a wilderness trip before the birth of his second child. It was good to have him along. Not just for the rare chance to catch up, but also, as an experienced sailor, I knew Steve's guidance would be useful in some of the exposed stretches of water we were to cover.

We paddled slowly north, the evening light closing in. It was June, and the lochside was dense with foliage. Birch, oak and alder spilled from the banks, their leaves overhanging and touching the water. Where the trees gave way, we saw farmland. There were small white cottages, sheep being herded and the distant sound of quad bikes. After an hour we moved into more open water to ready for the crossing to the northern side of the loch. From shore to shore the distance was just under a kilometre.

It's an edgy place, being so far out from land. A canoe is a highly susceptible craft, vulnerable to even the slightest changes in weather conditions. Winds can arrive faster than the time it takes to renege on your course, and gusts that would hardly register your attention on land can quickly kick up swells capable of upending a boat. Open-water crossings therefore require caution and commitment.

Almost halfway out, from nowhere, we were hit by a sudden squall. The canoe spun around, the bow flicked clockwise like a weathervane. I powered hard on the leeward side to bring the boat back on course, but I was unable to counter the wind and felt the canoe being forced in the wrong direction. For a few moments I experienced a quiet, concentrated panic, wondering if we'd be pushed into the middle of the loch, then caught up in larger, cresting waves. But the squall died just as quickly as it arrived, the water flattening and the boat responding once more. It was, though – as we were to later find out – a demonstration of the energy the loch could instantly summon.

We reached calmer waters on the northern shore, its surface turning to glassy green. Near a wooded peninsula, a small, scrubby island came into view. It had a swampish look, hardly sitting above the level of the loch and colonised by water-loving trees; willow and alder threw out thin branches, breaking the island's outline so that it was hard to tell where its shoreline began. It was a crannog, one of dozens of prehistoric settlements built into the body of the loch thousands of years ago. Remarkably little is known about these artificial island refuges, but most are believed to have been homesteads: family dwellings kept safe from animals and intruders by their distance from the land.

We scanned the treeline beyond the crannog for a place to make camp. It had begun to drizzle, and tiny droplets puckered the loch's surface, turning its texture from mirror to sandpaper. We found a spot, spaced between the canopies of two aged oaks. I rigged a tarp between them to provide some shelter from the rain, and we cooked dinner underneath. We were on the edge of a temperate woodland that stretched deeply inland behind us. The atmosphere was heavy and filled with the onion scent of wild garlic. Mosses grew everywhere, upholstering boulders and tree trunks in lustrous greens and yellows, while tresses of ghost-white lichens hung from every branch.

At first light the next day we explored the shoreline on foot.

Small boulders covered in black algae lay in between tufts of grass, making it difficult to walk any distance without slipping. Where the grass grew longer, flowers had taken hold: pink campion, forget-me-not, bluebell, and meadow buttercup. Up ahead, Steve signalled from a break in the forestry. There was a ruin, fifty metres from the water's edge, hidden among the trees.

Only the walls remained, spanning the height of two floors. I stepped across the threshold; inside, but still outdoors. A huge sycamore had grown from the centre of the building, its canopy reaching through the roof's empty space, dappling sunlight on the walls. The building was a drystone construction – the tallest I'd ever come across and almost a foot deep in parts. The stones were a myriad of shapes and sizes, but so perfectly tessellated that in places they appeared as a single rockface, fissured laterally with cracks. I could sense the builders' patience and expertise, the contemplation of placement for every piece of the wall, and I struggled to remember seeing anything more beautifully crafted.

Despite its size, the building was slowly disappearing, being swallowed back by the wood. Ivy streamed from the tops of the walls, sending down vines like rappelling ropes. In both rooms, fireweed grew in tall clusters, reaching for patches of sunlight, while spleenwort ferns emerged in splayed starfish forms from the ruin's stonework course. Outside, birch and alder saplings were establishing themselves. In between them, stands of foxgloves had sprouted, their bugle-shaped flowers turning from white to pink.

Back on the water, we pushed further up the loch, and after a couple of kilometres the islands appeared on the horizon: a small flotilla, fully rigged with dark-leaved sails. We paddled past more crannogs on the way and stopped at the largest, a circle of boulders about twelve feet in circumference. It was a desert island, a granite mound barely above the water and devoid of vegetation. A place to be marooned or abandoned on.

We headed to the furthest island in the archipelago first: a shoreless landmass of tall trees mirroring down into the loch and joined to the mainland by a causeway of wooden groynes. Being the nearest to land, it had signs of being frequently visited. A small stone pier tilted out from beneath the branches, and in the island's centre an area of flat ground held the remains of a wild camp: charred logs and a circle of smoke-blackened rocks. As we pulled away from the island, a large bird heaved itself into flight, its wings breaking in long, mechanical beats. It appeared black at first, darkened against the brightness of the sky. Then colour. Chocolate-brown wings and the flash of white underside, dark primary feathers and a distinctive highway-man's mask of feathers banding its face.

Ospreys were once all but extinct in Britain. Habitat loss, but, more devastatingly, human persecution, had gradually decimated their population and range until the last breeding pair was recorded in Scotland in 1916. By the early part of the twentieth century, it meant that the species had effectively been wiped from our shores. The only ospreys sighted for many decades afterwards were fleeting visitors, rare passage migrants normally blown off route on their journey to safer breeding grounds in Scandinavia.

The birds had initially been subjected to a similar fate as many other raptors – poisoned, shot or trapped for little more than sport or misguided hatred. The rise of the grouse moors and sporting estates in particular did much to hasten the decline of many birds of prey in the eighteenth and nineteen centuries – the employment of gamekeepers, coupled with functional refinements in firearms, meant that any animals which were perceived as a predatory threat to game were treated as vermin and shot on sight. Gamekeepers' log books from this period detail an annihilation of animals on estates across Scotland on a truly industrial scale. From a modern-day perspective, the quantity of birds killed is hard to comprehend. In his book *Days with the Golden Eagle*, the naturalist Seton Gordon cites the astounding number of golden eagles despatched on the Glen Garry estate in Knoydart between 1839 and 1840. In just two years, on this one estate, over ninety-eight were recorded as having been destroyed.

In theory, ospreys should have been of less interest to ghillies and sheep farmers. The birds' exclusive diet of fish meant that they were technically of no menace to estates' ground-nesting birds, and they would have no cause to take lambs – as eagles were frequently accused of doing. They were, though, easy targets. Ospreys are birds of habit, returning year after year to the same nesting sites, repeating the same patterns of territorial behaviour that would have made them simple to locate. During the breeding season they are also protective of their nests, circling low in the sky above their roost when they sense danger and therefore easily fixed upon through the crosshairs of a rifle.

Ironically, however, it was the birds' increasing rarity that became their downfall. As their numbers declined, the species fell into a rapid spiral towards extinction, driven there by a group of individuals who were both compulsive and resourceful in their ambitions.

It was egg collectors who were to cause the eventual demise of the osprey in Britain. Nineteenth-century oologists (the predominantly amateur band of natural historians that took to studying birds' eggs and nesting sites) approached their subject with all the scientific zeal and domineering indifference that befitted a Victorian gentleman. No nesting site was too remote, or breed of bird off limits. In fact, the rarer and harder to obtain, the better.

For most, it amounted to an intoxicating hobby. As well as being compelled by genuine ornithological enquiry, there was also the thrill of the chase, the endeavour and adventure of locating, then raiding, a perilous eyrie. And, of course, there was the prize. The kudos that came with pulling out a display drawer neatly arranged with a hoard of the rarest eggs, to the delight and admiring murmurings of your fellow oologists.

Some of the most extensive and detailed writing about egg collecting at this time comes from one of its most prolific practitioners, Charles St John. In his book *A Tour in Sutherland*, published in 1849, St John and his companion (the equally rapacious collector, William Dunbar) hunt, fish and shoot their way across the north of Scotland with as many moral contradictions as they could muster. Having already shot and dissected a rare red-necked phalarope on the shores of Loch Naver, killed a greylag goose on Loch Meadie to prove to his friends that it did indeed nest there, and attempted to bring down a pair of golden eagles, St John is then tipped off about an osprey nest on an island in a loch near the village of Scourie.

St John immediately sets off and discovers the nest, 'on the summit of a rock about eight feet high, shaped like a truncated cone, and standing exposed and alone in the loch' – a place, were it not for the attentions of the hunters, that would have ensured relative safety for the breeding birds.

A plan is hatched for St John's colleagues to approach the eyrie and nesting female by boat, while he lies hidden on the loch side, so that he 'might have the chance of shooting the old osprey herself in case she came within shot'. St John does indeed shoot the bird, which 'after wheeling blindly about for a few moments', he reports, 'fell to leeward of me, and down amongst the most precipitous and rocky part of the mountain, quite dead'.

Knowing the eventual fate of ospreys, of their close partnership during the breeding period and that they mate for life, the rest of the extract rings with an unintended, but desperate, poignancy:

> She was scarcely down behind the cliffs when I heard the cry of an osprey in quite a different direction, and on looking that way I saw the male bird flying up from a great distance. As he came nearer I could distinguish with my glass that he was carrying a fish in his claws. On approaching he redoubled his cries, probably expecting the well-known answer, or signal of gratitude, from his mate; but not hearing her, he flew on till he came immediately over the nest. I could plainly see him turning his head to the right and left, as if looking for her, and as if in astonishment at her unwonted absence . . . Not seeing her, he again ascended and flew to the other end of the lake . . . then flew away far out of sight, still keeping possession of the fish. He probably went to look for the female at some

> known and frequented haunt, as he flew rapidly off in a direct line. He soon, however, came over the lake again, and continued his flight to and fro and his loud cries for above an hour, still keeping the fish ready for his mate . . . As we came away we still observed the male bird unceasingly calling and seeking for his hen.

St John's colleagues return, delighted with their haul, in possession of 'two beautiful eggs . . . of a roundish shape; the colour white, with numerous spots and marks of a fine rich red brown' – the episode demonstrating not only the indiscriminate actions of the egg collectors, but also the scale of their impact. A threefold ecological travesty: the female shot, the eggs stolen and a breeding pairing ended.

It was not long before the effects of such a paradoxically venerating and destructive practice were to take their toll. And although by the turn of the twentieth century there was a growing belief that any scientific advancement from oology was far outweighed by its environmental costs, it was already too late for Britain's native ospreys.

From that point, the prospect of the birds ever returning to Britain seemed bleak. Ospreys have a highly ingrained philopatry instinct – the tendency for a species to return to the place it was born in order to breed. Once that link is broken, the cycle of birth and homecoming that binds a population and a place is almost always permanently lost.

Over the years, sporadic and uncoordinated attempts were made to tempt ospreys back to their old breeding haunts, but none were successful. Then, in the early 1950s, something remarkable happened. Increasing reports came through of ospreys being sighted in various parts of the

Highlands. At the time there was little to suggest that the birds had returned to breed until an article in *The Scotsman* newspaper in 1954 by the eminent ornithologist Desmond Nethersole-Thompson stated that a pair had, in fact, successfully bred at an undisclosed location on Speyside.

By now, several members of the Royal Society for the Protection of Birds and the Scottish Ornithologist Club had been spurred into action in a highly secretive mission. 'Operation Osprey' was launched the following year, a covert campaign, created and run by the RSPB's then secretary Philip Brown and conservationist George Waterston, aimed at protecting other potential osprey nesting sites. Egg theft remained the major threat to osprey breeding, so when word came through that an osprey pair had been spotted in the forest at Loch Garten, a small team of volunteers was mobilised to mount a round-the-clock guard on a potential eyrie.

The undertaking was arduous but organised with military precision. A handful of observers rotated individual watches on three-hour shifts, huddled in a makeshift hide within the forest. Resources and volunteers were also limited. The night-watch consisted of one person crouched on a packing crate peering through a hole in the hide's canvas screening, often in sub-zero temperatures, and vigilant at all times for any human activity in the darkness. There were no radio communications, and the only warning system for the watcher to alert fellow volunteers further back in the forest was by pulling a line of taut string rigged to the arm of a sleeping colleague. 'It was then', as Philip Brown recalled in his book *The Scottish Ospreys*, 'only a matter of seconds for the dozing party to slip his arm out of the loop and roll on to his feet. Torch in hand, he could be alongside the look-out within twenty or thirty seconds, booted and ready for action.'

The operation endured numerous disappointments in its first few years. On several occasions the nesting site was inspected and then spurned by visiting ospreys. And even when the eyrie was successfully used by a breeding pair in 1958, the nest was raided by egg collectors before the volunteers had the chance to chase them off. However, in 1959, after four years of set-backs, the pioneering conservation efforts were finally rewarded. In that summer, three chicks were safely reared at Loch Garten. The mission was a success, but word got out and rumours soon began to spread about the returning ospreys.

In a bold move, the volunteers decided to publicise the story and, instead of restricting access to the area, invited the public to also view the ospreys. The response was incredible. Within the following seven weeks, over 14,000 people arrived to get a glimpse of the rarest breeding birds in Britain, an average of 300 visitors per day. 'The ospreys', as Philip Brown later observed, 'were "on the map", attracting tourists to an area much in need of them, so that by local standards, conservation had become big business.'

Within fifteen years, the ethos of Operation Osprey and the public interest it had generated had extended to other nesting sites in Scotland, during which time over 140 ospreys were reared. The legacy today is even more significant, and there are now estimated to be between 200 and 250 breeding pairs in the UK, with a range that includes England and Wales as well as their stronghold in Scotland. More figuratively, though, ospreys have become totemic animals, symbols of recklessness and reprieve, an emblem of wildness almost lost.

For a couple of minutes, we watched the osprey on its flight path. It flew in a straight line, steadfast on a bearing, first above the dark woodlands on the loch's northern shore, then further inland until its silhouette became a speck, blinking out from view.

I noticed something else as we looked skyward. Fleets of cumulus clouds rolled slowly above us, their shadowed hulls backlit by the morning sun. Above them, more clouds: fast-moving clippers, zipping across at three times the speed, their broken shapes continually altering, forming new conglomerations as they travelled. It was a two-speed sky, a sign of atmospheric disturbance and an augur of the winds we were soon to experience.

The boat cut cleanly through the loch as we headed to Inishail, the largest of the islands. We approached its north-eastern edge looking for a landing spot, a cove or a beach to run the canoe ashore. A wall of dense foliage butted against the water, rainforest-darkness glimpsed inside. Three crows wheeled from the tree-tops, haggling and turning on each other with loud cackles. We paddled anticlockwise, rounding a small headland still thick with impenetrable forestry. I began to wonder if we would need to swim from one of the nearby islands in order to make landfall. Then, at its southern tip, a small clearing appeared between the trees, a thumbnail of sand curving in a miniature bay.

We landed on a narrow isthmus joining the main part of the island with a thin, birch-covered peninsula that stretched out into the loch like a seaside pier. Our beach was one of two. On the other side, twenty feet away, separated by a section of grassy, flat ground was another curve of sand and shallow water opening onto a larger, more sheltered inlet.

It was late morning, and with time to kill before we explored the island, we built a small fire in the pan, brewed tea, and talked. As we sat on the sand, a gust of wind suddenly ripped in from the loch, sending orange embers wriggling up from the flames. Minutes passed, but the wind remained, then an hour, and still no change. We looked back on the route we had taken earlier that morning. In the open water, less than a kilometre out, were white horses: the loch was roiling and spuming with creamy waves.

'Force four easily, maybe force five,' Steve concluded. 'Good sailing weather.' In such conditions there was no possibility of leaving the island. The canoe would be flooded as soon as we were clear of the shore. Realising we were stranded, we immediately succumbed to further idleness. For a couple of hours, we hardly strayed from the fire, only moving to add more fuel when the flames died back. We cooked sausages, then pasta, and brewed more tea. By early afternoon, with the wind still bending the trees, we stepped inside the curtain of dense woodland to scout the island.

Its dimensions encouraged exploration: large enough not to be able to see across it but small enough to circumnavigate in less than thirty minutes. We headed to its highest point, beating through chest-high thickets of bracken until we reached an area of tall trees and chutes of sunlight enclosed by a low wire fence.

We entered through a long metal gate, large enough to fit across a farm track. The space inside was free of undergrowth, and light fell on tree roots in soft gold patches.

Dotted between stands of pines and cedars were headstones of various sizes and ancientness. Some dated from the last century; others were much older, the inscriptions almost entirely worn back into the stone. At the far end, in a clearing of tall grass and bright sunshine, was a row of stone crosses. The largest was beautifully carved, a Celtic cross with interlocking patterns cut in relief. There was a ruin as well, shaded at the side of the graveyard. Moss had grown across the tops of the walls, and grasses had filled the inside of the small building.

Before travelling to the islands, I had read about this place. In the Middle Ages, the site was supposedly once the location of a small Cistercian nunnery. A chapel had also existed here, serving first a Catholic, then (after the Reformation) a Protestant congregation, who would make the journey across the water from the mainland to attend services. When a new parish church was built on the shores near Dalmally in 1773, the island ceased to be a place of worship and its buildings fell into ruin. The burial ground, though, continued to be used, the land retaining its importance as a place of interment.

The antiquity of this connection was humbling. For more than a millennium, people had brought their dead to be buried here; had felt something sacred about the island and this spot. It felt like a history of the living, not the dead. A veneration of place as much as people. The graveyard was still somewhere visited, a fixture in the landscape that even now obliged remembrance. On some of the graves, flowers had been recently placed, their leaves and stems shrivelled, colours becoming bleached. By another memorial – a small wooden cross, rising barely a foot – someone had left an empty whisky bottle, propped companionably at its side.

Before leaving, we came across a stone tablet, embedded in the ground and partly covered in moss and pine needles. Several figures were depicted on the stone. In the centre was a representation of Christ on the cross, flanked by the carved image of a woman, arms aloft and holding up a chalice. To the sides I could make out what appeared to be knights, dressed in armour and carrying weapons, although it was hard to tell if they were held in protection or aggression. The work was incredible. After so many years of exposure to the weather, the tableau had become crude and ill-defined, but the artist's original scene remained, full of drama and devotion.

Moving across the rest of the island was hard work. We walked its edge, slipping and sliding on a mulchy understorey and clambering over fallen trunks, the water flashing brightly below us. The dense woodland of the island was a relatively recent feature. In 1859 the English etcher and art critic Philip Gilbert Hamerton described Inishail as 'a long green pasture in the middle of Loch Awe, of a very tame and quiet aspect, broken only by one rocky eminence, crowned with a few straggling firs'. Hamerton had chosen the island as the venue for a strange, self-imposed artistic exile, brooding dramatically that he could not 'settle down as Constable did, to paint subjects of inferior interest, when an inexhaustible land of pictures lies to the north'.

Hamerton made camp with suitable expeditionary pomp, erecting three 'huts' on the grassy high ground of the island. One was for himself, another was for his hired servant and a third, 'an old Crimean tent', was used for cooking and storage. There he stayed over the course of the summer into autumn, attempting to live out a Crusoe-like existence, sketching, composing poetry and swimming in the island's small bay, floating on his back to keep his cigar lit. For added effect, Hamerton felt it was his duty to educate his domestic retainer, teaching the unfortunate local farm-worker (whom he insisted on calling Thursday) to speak 'pure English' and beating him relentlessly when he didn't.

Later that evening, the windspeed dropped by several knots, shifting direction and casting our end of the island in wind shadow. I decided to venture out in the canoe alone, to assess a likely crossing back, but also to paddle in the channels of the small archipelago.

Lying directly south of Inishail were the piratically named Black Islands. I crossed the short stretch of water to the largest of them, sculling the canoe slowly alongside its rocky shore. The place looked almost fictionally wild: lofty pine, the relics of the Caledonian Forest, looming over tiny shingle beaches recessed secretly in tree-covered bays. There was little doubt the islands had barely changed in hundreds of years, harbouring an ancientness that seemed both forbidding and enticing. During his time on the loch, Hamerton had expressed a similar astonishment. 'In natural beauty, both in shape and vegetation,' he wrote, 'they are the finest on the lake. To glide through the narrow strait on a summer's night and see the moon moving through the trees . . . was a favourite amusement of mine.'

I edged cautiously out to the furthest island, staying safely in its lee. The sun was low and bright in the western sky, sending its reflection in a glimmer path across the loch. Closest to me it was a thin trail, but further out the light's path became distended and broken, signalling turbulent water. I turned the canoe around and headed back to our campsite, seeing the smoke from the campfire billowing out above the trees.

We stayed by the fire that night, watching its glow intensify as the evening darkened around us. I awoke at five o'clock the next morning and peered out of the tent. The light was loose and grainy. I rubbed my eyes into focus and saw the loch mirror calm. The previous day's gales had abruptly ended, leaving the island ringing with silence. We wasted no time in departing while we could, unsure of when conditions would change again. We packed the canoe and slid the boat out onto the steel-blue waters, paddling back in a cloudless dawn, our clothes still smelling of woodsmoke.

Several months later, I journeyed to another wild island, Eilean Fhianain – or St Finnan's Isle in its anglicised form. Rising as a small, grassy dome in the narrow crook of lower Loch Shiel, I had long heard of the island's geographical uniqueness and its numinous reputation. After the seventh century, when St Finnan established a cell there, it had become a place of fervent spiritual convergence, a nexus point of devout journeys made from across the west of Scotland and beyond. All around the island there were signs of this religious magnetism. Pilgrim paths, coffin roads and marker cairns seamed from all directions towards the isle – radials of faith and devotion, a meshwork of belief etched in the landscape.

I followed the line of one of these ancient pathways, tracking its course by canoe along the northern shore of the loch. I had the wind at my back as I paddled: big blousy gusts that knocked flickering yellow leaves from beech trees and powered me through the water. Waves carried me too; the loch's seven-kilometre-long southern fetch had whipped up corrugated sets of whitecaps that, if I angled the boat in exactly the right direction, I was able to ride for several seconds at a time.

It was exhilarating, and before I knew it St Finnan's Isle was ahead of me, a bulbous plug of almost treeless land with two small channels running either side of it. On the top I could see monuments, the outlines of numerous stone crosses sprouting at different angles from the crest of the island's summit. Even from afar, it was clear that it was a place of spiritual significance, hallowed ground.

The customs and occurrences of island burial are widespread in the wilder regions of Scotland and Ireland. The relative abundance and accessibility of so many islands, distinct from main areas of habitation, meant that they were an obvious choice for the interment of the dead. From a practical perspective as well, they were beyond easy reach of many wild animals. Wolves, in particular, still roamed Scotland as recently as the eighteenth century and were often believed to be responsible for disturbing graves: 'Thus every grave we dug / The hungry wolf uptore, / And every morn the sod / Was strewn with bones and gore', wrote Eliza Ann Harris Dick Ogilvy in *A Book of Highland Minstrelsy*.

There is also a deeper, less rational reasoning that explains many of Scotland's island graveyards, rooted in a sanctity which is perhaps easily conferred on island spaces. Throughout history and across cultures, islomania in its most mystical and metaphorical forms has rendered islands with attributes of powerful spiritual significance. They are places physically and notionally abstracted from everyday life, mythologised territories such as the island of Avalon, where King Arthur is taken to recover from battle and where Excalibur is forged, or idealised sanctuaries such as Thomas More's fictional island of Utopia. Journeys to actual islands, then, become tacitly analogous, the crossing of one dimension to another, voyages that are both explicitly real and implicitly metaphysical. It's easy to imagine the symbolism that would inhabit and could be imposed on such detached wild spaces by both Christian and pre-Christian Celts.

As well as Inishail, there are several other burial islands that I knew of first-hand, each as intriguing as they are solemn. There is the beautiful and densely wooded island of Inchcailloch, in Loch Lomond – once the home to the Irish anchorite St Kentigerna, the resting place of the MacFarlane and the Macgregor clans, and where, every spring, thousands of bluebells grow in a blur of mauve, misting the forest floor. There is also Handa, the desolate island off the coast of Sutherland, which for centuries had been used to bury the dead from the mainland, supposedly to avoid the ravages of wolves once abundant in the region. In the narrow channel of Loch Leven near to the village of Ballachulish there is Eilean Munde, where headstones fashioned from the local slate quarry protrude between swathes of tall grass and pine trees.

Most beguiling of all is Isle Maree, in the vast northern loch of the same name. The tiny island is yet another place of ancient hermitage and religious mysticism. First settled by St Maol Rubha in the eighth century, the isle had later been reputed to exert unexplained healing powers. Local tradition suggests that to be immersed in the waters by the island, and to drink from its well, can provide a cure for madness. Any visit to the island, however, is bound by unequivocal superstition. Legend has it that nothing is to be removed from its shores, lest madness (instead of being cured) be invoked. It is a belief, I had heard, that is still observed by forestry workers whenever they visit the island.

I had once tried to reach the place, but had been beached by the wind on a nearby, nameless island. Unable to paddle my canoe any further, I made camp on a small strip of sand which was impressed by animal and bird prints. The next morning, I could see Isle Maree in the distance, a strange tumult of deciduous canopies, starkly different from all the other pine-clad islands in the loch. Knowing of its legends, and sensing its uniqueness, I felt almost relieved that I would be unable to complete the journey.

To reach there, though, is to be in a place of deeply stated but unspecifiable importance. As well as the remains of a Christian chapel, there is evidence of pre-Christian, pagan worship, and Viking graves are said to be among those in the burial ground. Near the site of the holy well is a wishing tree: the bough of an oak implanted with dozens of ancient coins, their copper oxidised to the blue-green colour of bright seawater.

Eilean Fhianain is an island necropolis. In every direction I looked, there were graves. From the small stone slipway on the northern edge of the island, I walked in between all manner of burial markers. Stone crosses, headstones, and tablets. There were small metal plaques that reminded me of the kind found in memorial gardens, and large slabs embedded in the turf: shiny black granite with bright gold font that seemed to expand as the light caught it. I found a wooden Celtic cross, part submerged in the ground. Only the top half of it was visible, as if the land had closed in on it, swallowing it up like a snow drift.

On the lower part of the island, two plots had been freshly dug; the earth still grey, sods of grass placed across the top. Next to one was a bouquet of bright roses, pinks, yellows, and whites, held together with a glossy bow. On the other were green foam blocks, stalks still sticking out, the flowers long since withered away.

I moved up the short slope to the higher ground on the eastern side. From the top I could see the full extent of the island. From edge to edge, it was only 200 metres in diameter, its shape squat and circular in the water. Beyond it, the channel dog-legged in a tight chicane, bending hard left, then hard right, so that the island sat orb-like, ball in socket at the loch's joint.

Several large structures crowded the summit. Two crosses dominated the skyline, one built on a tall stone column facing out onto the loch, the other vying for attention on a seven-foot-high frustum of perfectly shaped granite bricks. Next to that was a small sepulchre. Its roof had long vanished, but most of the building remained, and through the small doorway I could see several headstones against its far wall.

Occupying the largest area on the hillock was the outline of a large chapel, fifteen feet wide and extending lengthways for almost seventy feet. The walls were in ruins, a talus of masonry scattered on the ground outside. A doorway was still visible though in the south-west corner, and out of subconscious reverence, I walked round the edge of the building to enter there.

I stood at what would have been the back of the chapel, looking down the narrow reach of the nave. On the two longer walls, I noticed several large window recesses spaced along the length of the building, three on each side. The building must have once been filled with immense light and a view looking out across the water towards forest, hill and moor. And what a sight it must have been, this huge longboat of a building, perched high on such a small island.

A chapel is believed to have existed here from the time of St Finnan, its final iteration only eventually falling into ruin in the seventeenth century. Its position in the vast watery domain of Loch Shiel made perfect sense. In a time before roads, when journeys by boat were still the most efficient form of travel, it would have provided an ideal venue for summoning a congregation. But even before he began celebrating mass there, St Finnan must have also realised the advantages of claiming an island base.

The cleric was a missionary, a contemporary of St Columba and part of the great Hiberno-Scottish spread of Christianity. As such, he knew that geographical reach was everything. Distances covered equated to minds converted, and Scotland's lochs and seaboard were the perfect conduits for the dissemination of ideas and teachings. Loch Shiel's 28-kilometre-long shore, in particular, allowed access to a vast area of land, from Arisaig and Morar at its northern end to Sunart and Ardnamurchan in the south.

Christianity swept rapidly through this part of Scotland, modifying and amending ancient practices of Celtic paganism rather than displacing them entirely – a new faith built on the old. For the next 700 years, Moidart and the lands west of Loch Shiel remained deeply imbued with Catholicism, closely linked with the Irish monasteries as well as the papacy of Rome. This, however, was to decisively change during the sixteenth century. As the religious firestorm of the Reformation raged across Europe, Scotland also became engulfed in its flames, and in 1560 the Scottish Parliament passed a series of Acts that outlawed the Roman Church in Scotland. Across most of the country, Catholicism was forced to become an underground faith, its sacraments celebrated in secret to avoid persecution and even death.

Much of the West Highlands, though – and Moidart especially – was left in a state of religious limbo. Churches were requisitioned for Presbyterian worship, but clan chiefs would often provide their tacit protection to remote Catholic populations. The result was a demography of divided faith. A spiritual no-man's land where Catholicism was neither fully oppressed nor Protestantism fully proselytised; an uneasy but functioning coexistence.

The effects of this religious schism were evident on Eilean Fhianain. I had read somewhere that burials here had observed a specific ritual of placement – Catholics interred in the northern half of the island, Protestants in the south, the small island echoing the divisions of life in death.

At the far end of the ruin, a rectangular slab of granite rested across two stonework columns forming a small altar. Behind it, a foot-high niche held a cruciform stone, its surface neatly incised with the outline of a cross. The fixtures made the shell of the chapel look oddly serviceable, as though the only thing missing was the priest and his congregation. Most improbable of all, though, was an object resting at one end of the altar: a hand-bell, quadrangular in shape and roughly six inches high, the metal weathered to deep shades of jade and black.

Bells of this kind were characteristic of the early Irish Church, their use closely associated with the liturgical traditions of the wandering monks and missionaries like St Finnan. I had seen others like it before: religious artefacts tucked behind glass display cases in museums, information plaques next to them noting their importance as evidence of the dispersion of Christianity. That such a relic could be found here, unguarded yet in the same place for so long, was astounding. I struggled to comprehend why or how it had remained on the altar for centuries.

Some months later I found reference to it in an obscure, out-of-print book. In *The Peat-Fire Flame*, published in 1937, the author, Alasdair Alpin MacGregor, devotes a chapter to what he referred to as bell-lore. 'Records of sacred bells are scanty,' he wrote. 'But it is well known that in certain parts of Scotland there existed bells, to which were ascribed healing and protecting properties . . . The general belief was that such bells, when stolen or otherwise removed from their rightful place, had the power of returning of their own accord.'

MacGregor's assertion of some kind of mystical homing properties for these objects felt a tad far-fetched. Nevertheless, utterly intrigued, I read on. A few pages later, there it was. A section about Eilean Fhianain and its bell. 'The altar of the chapel on this isle is still preserved;' he noted, 'and on this altar for nearly three centuries has rested St. Finnan's Bell'. After the description, MacGregor detailed something else. A particular legend associated with the bell which seemed to explain its continued presence on the island. According to MacGregor, during Bonnie Prince

Charlie's flight to safety after the Battle of Culloden, government soldiers on the hunt for the Young Pretender searched Loch Shiel for him. On a sweep of St Finnan's Isle one of the Redcoats came across the bell and unwisely decided he would take it. Suddenly, the bell began to screech so strangely the soldier was compelled to immediately return it.

The story must have spread widely because apparently no one since has attempted to remove the bell, local lore upholding that a curse will surely befall anyone who tries to take it.[*] It seemed to me a curious dynamic. A wild history protected, maintained. Not the ghostly reverberations of a place, travelling and diminishing like light waves through time, but a living consciousness, perpetuated, enriched, existing at an intersection of memory and landscape.

* Less than a year after I visited the island, it was reported that St Finnan's Bell had gone missing. Sadly, at some point in that following summer, the bell was stolen. Its whereabouts remain a mystery, but if folklore holds true, it may yet somehow find its way back to the island.

Land of the Left Behind

The Atlantic Wall, Sheriffmuir, Clackmannanshire and *the Ardnamurchan Peninsula*

The thought stayed with me. In the weeks following my return from St Finnan's Isle, I rolled it around in my subconscious, turning it over again and again as if rubbing pebbles in my palm. The idea that wild histories were somehow synchronous with the present, influencing and occupying the here and now as much as they were signifiers of the past, threw a different perspective on the places I had visited. They had felt wild certainly, but the stories I'd encountered had contemporary significance, were sustained as well as sustaining. The ancient pathway still used on Jock's Road, the placement of flowers at Inishail's graveyard, the legend of St Finnan's Bell – these were not mere rites of remembrance but a vital continuation of what had been.

At the same time, a narrative was also building in my journeys. An observation of things left behind: memorials, shelters, gravestones. Things left behind – but not discarded. There was a subtle difference, an intentionality and motive that had endowed certain locations or landscapes with an enduring meaning. The opposite, however, was true as well. Markers exist within Scotland's wild places that are discrete and fragmentary, unconnected and now forgotten. In many ways these unremembered histories hold more intrigue, oblige more interest, require more explanation. They are the evidence of specific moments, sudden events and abrupt endings. Wildness is part of their story, but also the reason for their lingering anonymity. To find such a relic is therefore to be confounded, puzzled by opposing characteristics of significance and obscurity. One such anomaly in the landscape astonishes me more than most, and I visited it a month after my trip to Loch Shiel.

On the high, empty moorland of Sheriffmuir west of the Ochil Hills is a long section of wall, two metres thick and broken by artillery fire. It's not easy to see at first. The concrete takes the same dun colour of the surrounding landscape, a swell of hillside rising behind it, nudging it below the horizon. But it's there on the map, a thin, rectangular box sitting snugly, but without explanation, between the contour lines.

I parked on the single-track road that runs across the moor's western flank, opening the car door onto a wind that had all the force of an incoming wave. My sons were with me and bundled out of the car too, half falling, half jumping, one on top of the other. They raced on ahead, old enough to be let loose but still not tall enough to stride easily through the heather. I watched them leap and stumble across the moor in their bright rain jackets, heads bobbing above the tall grass. By the time I reached the wall, they were already there, chasing each other round the structure.

On one side, the wall ran straight for about 250 feet, a solid unbroken façade supported by several sturdy buttresses. On the other, the structure had been disembowelled. Sinuous metal peeled out from the broken concrete, thick cables splayed in the flower patterns of explosion. At one section, the wall had been completely breached by shell blast: a half-circle gap filled with a blue and white sky, concrete held in mid-air by wires, suspended at the moment of detonation. In between, grasses had grown, large tussocks with white seed-heads, covering the debris and joining the moor on either side.

It served no obvious purpose, this mammoth lump of industrial masonry. There was nothing in the acres of wind-swept moorland to protect or enclose, no architectural clues or logic to its placement. It simply began and ended. But it was here, in the secrecy rendered by a wild and remote place, that a crucial stage of Britain's war effort against Nazi Germany had been finalised.

The wall had been built in 1943 as a full-scale replica of a section of Hitler's Atlantic defences, the vast series of coastal fortifications that ran all the way from northern Norway to southern Spain. Plans stolen by the French Resistance, combined with aerial photography, enabled the British Army to build their own versions of the wall according to the exact specifications of the German construction. At least four other walls were also covertly made, one in Wales and another three in England.

The intention was to test the structural resolve of the concrete ramparts ahead of the Allied invasion of Europe on D-Day. Ordnance was heaped on the replica wall in intensive set-piece rehearsals, while soldiers from artillery regiments carried out mock offensives of the beach landings. The wall fulfilled its purpose. On the one hand it confirmed to the military tacticians that the German defences could be sufficiently breached to allow tanks, troops and armaments to pass through and press the offensive inland. On the other, it enabled a vital pre-visualisation, a means for soldiers to somehow steel themselves for the eventual horror of the beach landings. Amid the hail of gunfire and shrapnel as they ran and fell, sodden, bloodied and terrified on Normandy sand, a surge of familiarity might have surfaced in their thoughts; the muscle memory of the training, the self-belief borne of repeated action, the recollection of the wall, its concrete shattered once before, a battle theoretically won high on that heather-clad moor.

The peculiar fate of the Atlantic Wall, an object both unaltered and forgotten because of its wild location, put me in mind of another place that held secrets on a far greater scale. It was an entire region of the left behind, a formaldehyde jar of land that had preserved life as it had once been lived: sheilings, crofts and whole townships abandoned to the elements.

Jutting out into the daisy-chain archipelago of the Inner Hebrides, the Ardnamurchan peninsula is land among islands. The Isle of Mull lies directly opposite its southern shores, detached by only a few kilometres of sea. To the north the outlines of the Small Isles are visible: Rum, Eigg, Muck and Canna loom like battleships ready for invasion. To the west are the turquoise-fringed islands of Coll and Tiree.

Water surrounds and defines the boundaries of Ardnamurchan's landscape. Pastry-cut coastlines offer up frets and fingers of outcrop and bluff, hiding bright cuticles of grey-white beaches unreached by roads. On its only landward side, the River Shiel almost completes the hydrographic separation, diagonally crossing, but not fully severing its attachment to the mainland. It is also mainland Britain's most westerly point and notoriously difficult to reach: a panhandle of volcanic rock pinioned to the west coast, only joined to the rest of the country by a solitary, sweeping single-track road.

As such, it feels like a place of rarefied geography, its characteristics of isolation and inaccessibility prompting both the desire for occupation and the need to escape. Its human history has accordingly catalogued the influx of multiple peoples. Early Mesolithic nomads were among the first to arrive, scavenging the shoreline over 6,000 years ago before the customs of Neolithic tribes then took root: a settling of the land, farming and the burying of the dead. Beaker Folk – associated with the advent of the Bronze Age – were to come next, bringing with them new technologies for pottery and metal work and integrating with the established Neolithic populations.

The Bronze Age melded into the Iron Age, and around this time settlers from continental Europe – the Celts – advanced as far across as the Ardnamurchan peninsula. Their tribal hierarchies based on extended family units led by a clan chieftain were to form the main societal structure in the region for almost the next two millennia. There were others of course, Dalriada Gaels from Ireland and Norse raiding parties, arriving in waves from the waves – conquerors and colonisers, warriors of the sea who understood the strategic value of the surf-scraped headland as a place to overlord the islands of Scotland's west coast.

Ordered in this way, the successive cultures appear separate and distinct, one replacing the other; a moving in and a moving out. But in reality, each augmented its predecessor, adapting rather than eviscerating their forerunners' existence. It is a history which speaks of assimilation and change, acceptance and adjustment, what anthropologists refer to as acculturation – a process of gradual social modification, a blending of customs and habits which in turn creates a new, entirely unique set of cultural characteristics.

This layering of human settlement, the calcifying process of time, hardened and fixed connections with the land. Ardnamurchan and its peoples became coupled through hardship; consecutive generations learning to ply something more than just survival from the asperities of their environment. It's a relationship evidenced across the peninsula, visible in ruins of duns and hillforts, glimpsed in the outlines of stone circles and observed in the many cairns and burial mounds. All describe an insoluble link, an involuntary possession, a land that could claim ownership as much as it could claim to be owned.

I decided to travel to Ardnamurchan not only to see evidence of these landscape bonds but also to find a place where this connection had finally been broken, where the centuries of a collective past had been forcibly and irrevocably left behind. While I was there, I also wanted to follow the lead of another wild history.

In the weeks before my trip, I had come across the photographs of a local amateur historian. They showed the remains of a small, single-room building, the low, roofless walls covered in moss and bracken, hidden from view in the gully of a small waterfall. It was, I found out in an email conversation with him, most probably an illicit still – once the secret location for illegal whisky production. I had never come across one before, for none were meant to be discovered anyway, and now few still exist. They are, though, indications of tantalisingly subversive times, a period when Scotland's wild regions were alive with the clandestine industry of bootlegging.

The historian had provided me with a grid reference – six digits with which to locate it. In theory, the numbers would narrow my search considerably, being precise to a space of 100 square metres. But allowing for inaccuracies in either his reckoning or my navigation, the area could extend to 300 metres or more. Perhaps acknowledging this, and mindful of the concealing power of the landscape, the historian had also detailed features of the exact spot; 'set against a steep cliff', he had noted, 'in a gorge below a waterfall, with a burn running past'. It sounded idyllic.

I started from the middle of Ardnamurchan's vast moorland interior, pulling the car off the narrow road onto soft, peaty ground. My plan was first to walk north, towards the apex of the moor, searching for the still. Then I would travel east and south to reach the coast, aiming for a small bay, clustered on my map with the notations of ancient monuments. Finally, I would then circle back inland to find the deserted village of Bourblaige, which sat somewhere on the slopes of the ancient volcano of Ben Hiant.

It was late autumn, and the moor's summer colours had rusted back to the earthy hues of a water-colourist's palette: fawn, umber, ochre. I stepped through briars of dead bracken, fronds and stalks shrivelled to tawny plant skeletons that crunched underfoot. The wind was briny, blowing westerly, smelling of the shoreline. To the south, the sea was a luminous band, almost colourless: silver foil and, where the sun broke through, gold leaf.

The light changed dramatically on the moor. Roving cloud shadows brought the feeling of early evening darkness, of the day's brightness draining from the landscape. Sunlight was equally accentuating. As if by the flicking of a switch whole sections of the moor suddenly became flushed with detail and tonal depth. Within seconds the place could appear with startling clarity, only then to once again vanish within itself, a multitude of swells and dips merging to assume one indivisible mass. It was a place to become unseen, a landscape easily encrypted. For this reason, Ardnamurchan's moor (and much of the Highlands) offered the perfect backdrop for the trade of the illicit distillers.

By the second half of the eighteenth century, whisky had surged in popularity, shaking off its perception as a provincial liquor to garner something of a mass-market appeal. Much of the attraction was in its cost and availability. For many years, beer and ale had been taxed heavily, making whisky a far cheaper alternative. Meanwhile, with the advent of the Napoleonic Wars, tariffs and a patriotic opposition to French spirits meant brandy and claret – the drinks of the social elite – were also being replaced by whisky. At the same time, rapid population growth and increasing urbanisation were fuelling even greater demand for the spirit's consumption.

Whisky production became a highly profitable but increasingly taxed enterprise. Excise duty on the ingredients, manufacture and sale of the drink had been raised repeatedly to help fund Britain's costly war with France, turning licensed distilling into a low-margin commercial activity. As a consequence, and despite attempts to prohibit its practice, small-scale illegal whisky distilling flourished. In 1782 over 1,000 illicit stills were recorded as being seized by the authorities. According to the eminent historian Tom Devine, this number represents only a small fraction of the stills which remained active and undetected across the north of Scotland, a figure he conservatively estimated at twenty times that amount.

The huge volumes of the illegal whisky trade were driven by more than just the combined efforts of unconnected, opportunistic individualism. Collusion was vast and existed at every stratum of Highland society. So much so, Devine wrote, that 'lawful authority was defied on a remarkable scale'. The actual stills were worked on by the largely agrarian population as an invaluable supplement to their meagre incomes. Men, women and children were all involved in the enterprise, from the transporting, washing and malting of the grain to the cutting of peat for fuelling the distilling process, or as look-outs and messengers for the whisky-makers.

The land-owning class were also implicitly involved in enabling this cottage industry. Tacit approval for illegal distilling was often given merely by turning a blind eye to their tenants' bootlegging activities. However, more overt encouragement was also evident through the lack of legal punishment. Many landowners were establishment figures and often fulfilled roles as Justices of the Peace, dispensing lenient judgments to those caught producing or smuggling whisky. In 1819 more than a quarter of the 4,201 such cases ended up being dismissed by the local magistrate, perhaps through an acknowledgement that any interference in the distilling practices would jeopardise their tenants' rental payments.

Such was the importance of illegal distilling to many of the economically impoverished areas of the Highlands that it was often the only difference between subsistence and poverty. Population numbers in remote areas were sustained and even increased during periods of unfettered whisky production, but were equally depleted by more rigorous enforcement of excise laws. Twenty years after the Excise Act of 1823, the *Statistical Account of Aberdeenshire* (1843) recorded effects of the legislation which were typical for many Highland glens: 'While this infamous and demoralizing practice prevailed, populations increased through the facilities by which families were maintained among the hills and valleys on its profits. But no sooner was this system put down, than the effect appeared on population. Fewer marriages than formerly now take place, and a considerable number of families, formerly supported by illicit distillation, have been obliged to remove to towns and other parishes: a good many families, also, have emigrated to America.'

Few other areas of Britain could have been better predisposed to the illegal production of liquor. It was, as Devine observed, an activity 'well-suited to the socio-economic communalism of the Highlands. Social links at a local level preserved that fellow feeling against the law so vital to the success of the illicit trade'.

In a landscape too where agricultural output was challenged, producing whisky was a remarkably feasible process compared with forging other livings from the land. The equipment for the illicit stills was reasonably easy to fashion, requiring little investment. Material inputs were also easy to come by. Grain could be obtained with relatively little cost, and peat and water were in abundance. Perhaps most importantly, the distilling process was not dependent on the fickle contingencies of the Scottish climate. Production would not be significantly impacted by weather conditions, and even for the outdoor stills a basic form of shelter could be made to ensure operations continued year-round. Once constructed, these stills would also be incredibly difficult to spot. Hidden in remote crevices the structures were assembled from the natural materials that could be found nearby: small buildings of heather, bracken, wood and stone that would blend perfectly with their surroundings.

The moor held onto its secrets carefully. I walked for an hour or so, surprised by how subtly the landscape revealed itself; small details only given up within the space a dozen footsteps. From a distance, what I thought were broad expanses of flat ground turned out to be riven with culverts and gullies, hidden watercourses that were recessed below eye level, heard before they were seen. I crossed several of these sunken channels, scrambling down into them clutching at the woody roots of heather to anchor and slow my descent. Once in their hollow, I too disappeared from view, momentarily invisible, feeling like an animal running to ground.

As I neared the historian's coordinates, I decided to stay within the concave dip of one of the burns and follow it upstream, hoping it would lead me to the waterfall and the still. The water crooked and kinked, looping back on itself, scurrying behind corners and turning mazy right angles. From the banks above, grouse repeatedly exploded against the skyline, bursts of feather and sound that startled me every time.

The burn divided in two, and I chose left. It thinned again at another V-shaped confluence, and I picked left again, half expecting to soon be doubling back on myself. Remarkably, though, the land started to rise ahead of me, the gully turning into a small ravine, alder and birch saplings sprouting from the sides. I had reached a headwall, and the stream now tumbled white over black rocks, a waterfall cleaved into the dark underside of the moor.

Next to the falling water, tucked into its cleft and therefore inaccessible to deer, a large tree had grown, branches reaching out sideways from the crags. It was strange to see such a profusion of growth on the moor. Nothing larger than the wind-stumped heather is normally able to flourish in such a place. Hostile weather conditions, acidic soil and the bark-loving appetites of animals halted the growth of any shrubs before they could begin. But here, the waterfall and the small cliff offered a rare sanctuary. It seemed logical that the still should also be tucked away in this place of refuge, and I saw it on the opposite bank.

There were three walls, rising a couple of feet off the ground and shaped in a neat rectangle which butted against the side of a small crag. The construction looked incredibly solid: boulders stacked almost a foot deep, easily strong enough to support a roof propped against the bluff on its fourth side. A layer of turf had grown on the tops of the wall, securing the structure like copping stones. The entrance was at the far end, a gap in the blockwork facing into the cleft and the waterfall.

It felt an intuitively odd point to position access, but made perfect sense. Surrounded by miles of uninterrupted darkness, at night the firelight from the small bothy would be contained within the culvert of the waterfall, ensuring the bootleggers' location would not be inadvertently revealed. Between the entrance and the burn another wall had been built, a small grassy ramp which would have protected the building from the stream bursting its banks. A notch had been left in the bank's far end, a recess big enough to allow pipes to run through, funnelling the cold water needed for the distillation process.

Soft reeds and bracken filled the area within the walls. Even abandoned it felt like a functional, practical space, ten feet long and four feet wide. The place could have easily accommodated all the paraphernalia necessary for whisky-making. Typical operations involved a large copper still containing fermented whisky mash heated over an open fire. The evaporated liquid would then rise through a coiled pipe (the worm), which was cooled with water from the nearby burn. As the vapour condensed on the cold metal of the worm, the distillate alcohol would then be collected in a separate container for further rounds of evaporation. The whole process was time-consuming and monotonous, requiring patience and attentiveness from the whisky makers. Because of this, bothies were often built not just to house the equipment, but to enable shelter for days, sometimes weeks, on end.

I scouted around in the building's undergrowth, searching hopefully for any clues or trinkets to somehow confirm its former use. There was nothing, but nonetheless it seemed impossible to ascribe it another purpose. The structure was too small to be a permanent dwelling and too awkwardly placed to be used for sheltering cattle. What's more, it was deliberately hidden, unconnected by paths and miles away from the nearest village. It felt perfectly in keeping with the nineteenth-century paintings of illicit stills I had seen: a shelter profiled into the landscape, a thatch of heather arching over a cramped interior, smoke filling the scene, figures silently concentrating, tending their task in the glow of firelight.

I climbed up out of the gully and once again became level with the moor. The wind hit me immediately, cold and salt-edged. Looking down from the lip of the waterfall, the structure had become almost invisible, its linear footprint dissolving into the land around it. Fully roofed, it must have been impossible to spot from above. Even for those who knew of its whereabouts, finding it would have proved extremely difficult. The only guaranteed method of location would be to track the burn as it reeled upstream, memorising the correct sequence of turns for each confluence: right, left, right, left, left. I liked the idea of this clandestine landscape knowledge, the land coded like a combination lock, held onto secretly in the minds of a few.

From the waterfall, onward navigation was easy. Behind me, to the north and west, a line of low hills pressed hard against the sky, outcrops bared like fists above the moor. I made my way slowly downhill heading for the coast, cutting between two large forestry plantations whose densely packed treelines extended like a perimeter fence for several kilometres. The woodland held an unnatural darkness, condensing and absorbing so much light that even the air above it had the appearance of dusk. Eventually I moved into open country once more, and as the slope increased I made sight of the small bay of Camas nan Geall.

A mottled-white beach, the colour of old snow, sat at the edge of a wide bowl of grassland forming a semi-circle facing south-westerly into the sea loch. The tide was out, and streaks of shiny wet sand threw back oily reflections below the wrack line. A stubby headland, edged with black boulders at its base, rose at the bay's eastern end, and on its western side a stream fell out from the moorland, cutting a wavy line across the shingle. The peninsula stretched further west and south, closing the bay in a pocket of windless water.

I reached the high ground above the cove, standing in a small, empty car park that kept an interpretation board set on a drystone plinth. The board faced directly onto Ben Hiant, proclaiming in large font: 'Volcanoes Ahead!'

The hill was hard to take in at first, its shape difficult to define. The summit and the ridgeline lurched and floundered, a series of lumps, gullies, mounds and ledges that made its outline wobble across the horizon as if caught in the quiver of heat shimmers. Its southern flanks bumped steeply seawards, pouring straight into the water in a confusing tumble of geological forms. Along its eastern face, the geography was equally elaborate, the hillside split by a diagonal terrace of exposed rock and a long shelf of flat ground that divided Ben Hiant into upper and lower realms. Captured on the map, the hill's representation was even more complex, a tangle of contour lines, swirling and looping in and around each other in a chaotic pattern of whirls and eddies. It was one of the most disordered hillscapes I had ever seen, a jumbled topography shaped by tumultuous events that began over sixty million years ago.

During the Palaeogene period, enormous tectonic forces were at work with vast landmasses being brought together. India was in the process of colliding with Asia, forming the Himalayas, while Africa was edging northwards to enclose what is now the Mediterranean. Elsewhere, continents were also being pulled apart. Australia had begun to detach from Antarctica, and America and Eurasia continued their separation, stretching the Atlantic Ocean and leaving a trail of volcanic cataclysm in their wake.

A ring of fire blazed the continental margins on both sides of the North Atlantic, from the Faeroe Islands and East Greenland to the north of Ireland and the north-west of Scotland. The mountains of Skye, Rum, Mull and Arran and the islands of Rockall, St Kilda and Ailsa Craig were all formed in these furnaces of molten rock, and Ardnamurchan owes almost all of its landform to this violent period in the Earth's geological history. Indeed, Ben Hiant's confused appearance and elevation were the direct results of this volcanic activity.

To the south, lava fields from eruptions that created the Isle of Mull flowed far enough to settle upon the much older sedimentary rocks of Moine schist that form the base of the hill. At the same time, to the west, a huge dome of magma was building beneath the Ardnamurchan volcano. Lava from the chamber oozed through vents in the schist, pooling and crystallising before it reached the surface. This fine-grained basalt and dolerite was later exposed though the harrowing forces of glaciation, forming the chaotic undulations seen on Ben Hiant's summit today.

I had not realised it as I searched for the still, but this volcanic mountain-building was even more significant inland on the high moorland of Ardnamurchan. The line of hills that I had seen edging the skyline, their exposed rockfaces looking like grazed knuckles, were, in fact, part of a huge volcanic rampart circling and enclosing most of the peninsula.

As with related landforms elsewhere in the world – calderas, sanctuaries and craters – its scale is so vast as to be almost unnoticeable at ground level. Only from the air can its impression on the landscape be properly registered. Viewed from above, by satellite or aeroplane, the western half of Ardnamurchan becomes imprinted with concentric circles, the largest over twenty kilometres in circumference.

The effect is both beautiful and hard to comprehend. The land takes on an almost liquid form: disturbed water captured in high-speed photography; a rock thrown in a lake or a raindrop falling in a puddle. The largest outer rim of hills appear momentarily in motion, rippling from its centre, surrounding a depression that looks set to level and disappear at any second. But the image and the landscape remain, a huge circular valley many kilometres wide, sunlight and shadow falling across it. Seen in this way, the sheer size of the Ardnamurchan volcano confounds perception.

The day's light was already beginning to fade as I stepped down into the bay. A vehicle track cut through the middle, all the way to the beach. On either side was pristine grassland, a huge paddock neatly fenced with lush green turf. There were no animals in it, and it looked extravagantly unused, like the grounds of a stately home lifted and placed within the rough confines of moor and sea.

In its centre, the track passed by a line of large sycamores whose lower branches drooped onto the raised ground of a Neolithic cairn. The mound was a burial chamber, about 6,000 years old, which had once extended far beyond the fenced area that now contained it. I pushed through a small wooden gate to get a closer look. It was still possible to make out the original structure. Two vertical stones described a small entrance, while behind it a large cap stone which would have roofed the cairn had slipped from its supports and now rested at a 45-degree angle. Most of the original stones were missing, but the ones that remained were blotched with grey and white lichen, camouflage patterns that resembled the mottled appearance of whale skin.

A ruin of a small rectangular building stood close by, and as I scanned the bay I saw several others. Less than a hundred metres away on a promontory above the burn was a small roofless croft. Opposite that was another building looking out onto the bay, its gabled walls and chimney still intact. And at the eastern end of the beach there was a stone cottage with trees growing inside. A tangle of branches filled the space where its roof would have been, so that the building looked as though it was sporting a fine head of hair.

Near the edge of the field there was a graveyard, marked on my map as Cladh Chiarain. The site faced out on to the crescent of beach, enclosed by a low wire fence. Inside, the grass was long and unkempt, with the summer's bracken curling brown at the sides of another small ruin. Several broken headstones lined one side of the wall, their carved reliefs softened by the elements.

At the shore end, what I thought was the marking of a large grave was actually a standing stone, dating from around 4,000 years ago. The monument was unusual. Not because of its shape or form, but because several thousand years after it had originally been erected, its symbolism had been appropriated – requisitioned by an incoming faith. On its seaward side, the stone had been inscribed with a large Latin cross, a smaller equal-armed cross and an animal with an upturned tail. The etchings are all early-Christian motifs, devotional ciphers made to mark a new spiritual assertion. This religious reclamation had also become permanently fixed in the name of the place, a reference to St Ciaran mac an t-Saeir – another Irish missionary who, like Columba and Finnan, had evangelised on Ardnamurchan's shores and is thought to be buried within the small plot.

I walked down onto the sand, stepping over bright debris in the tideline: green plastic bottle-tops, blue nylon rope, the yellow remains of a rubber glove. From the water's edge I turned and looked back. With distance and perspective, other features revealed themselves, further layers of human presence. I could see the infrastructure of a once thriving community – sheep folds, byres, even the stone-built causeway of a small jetty. It was hard not to feel the latent energy of the place, to sense a landscape still charged with memory.

Forgotten or forsaken places hold a potent enigmatic resonance in our collective consciousness. The trope of the lost city – a vanished civilisation existing somewhere beyond the visualisation of maps – occurs frequently in oral histories and in literature, and over the centuries has spawned countless real-life expeditions. Often, the imaginative pull of such apocryphal places has proved so powerful as to be deadly. Sixteenth-century explorers, including Francisco de Orellana, Gonzalo Pizarro and Sir Walter Raleigh, were so obsessed with locating the fabled city of El Dorado that they conducted numerous fraught and disaster-stricken expeditions to find it. Similarly, the British adventurer Percy Fawcett, utterly consumed with discovering the lost city of Z, ventured into the Amazon jungle, unsure of 'whether we get through and emerge again or leave our bones to rot in there'. His words proved prophetic and he was never seen again.

This fascination has also embedded itself culturally, appearing across a range of artistic forms. J.M.W. Turner produced two paintings of the ruined but once mighty port of Dunwich in Suffolk. In one, a group of men are depicted in the foreground straining either to launch or land a skiff. All around them, the sea is breaking on the shoreline. The water is almost entirely white, quiffed with extravagant, agitated brush stokes into a maelstrom of spume and spray. The men's struggle, the drama and the urgency of scene demands the viewer's initial attention, but the painting is dominated by its background detail. High on a prominent headland we see the shapes of skeletal buildings. Their outlines are indistinct, but Turner infuses the abandoned town with a strange luminosity, a spectral whiteness that lifts it from the canvas, making it appear both momentarily present and innately ephemeral, visible on the horizon but just about to fade from view.

John Constable was similarly drawn to another long-deserted but once powerful township and chose it as the subject of several of his works. He painted the hillfort of Old Sarum, first in oils in 1829, and then as a larger watercolour rendering five years later. The paintings, like Turner's, show a narrative scene: a lone shepherd moves his flock through empty fields below the grassy ramparts of the ancient settlement. Signs of habitation have all but disappeared, and the land has reverted to simple pastoralism as sheep graze underneath a glowering, cloud-laden sky. The story the paintings tell is of wildness restored, of nature overpowering the civilising urge of humans. No buildings remain (though we know that they were once there), and we are left to ponder the reasons for both the town's demise and the fate of its people.

In reality, the decline and abandonment of Old Sarum was gradual and undramatic. The site had been occupied from the Iron Age and was built upon by the Romans and the Normans before eventually falling into ruin in the fourteenth century. But Constable's artistic interest in Old Sarum and Turner's choice of Dunwich connect with a deep sense of lamentation rooted in the loss and desertion of hundreds of towns and villages across Britain. In the eighteenth-century poem 'The Deserted Village', the Irish playwright and poet Oliver Goldsmith offers up a nostalgic elegy for a lost rural idyll:

> How often have I loitered o'er thy green,
> Where humble happiness endeared each scene!
> How often have I paused on every charm,
> The sheltered cot, the cultivated farm,
> The never-failing brook, the busy mill,
> The decent church that topt the neighbouring hill

But the verses quickly switch from the wistful remembrance of a place, to the reality of a village depopulated and abandoned.

> Here as I take my solitary rounds,
> Amidst thy tangling walks, and ruined grounds,
> And, many a year elapsed, return to view
> Where once the cottage stood, the hawthorn grew

In doing so, the poem becomes polemic, an impassioned social commentary railing against the destructive, economic forces of 'trade's unfeeling train', which 'Usurp the land and dispossess the swain'. For W.H. Davies also, the imagery of the deserted village was equally symbolic and just as politically stirring. 'What tyrant starved the living out, and kept / Their dead in this deserted settlement?' wrote the

Welsh poet, obliquely referencing the devastating effects that enclosure, sheep-rearing, industrialisation and urbanisation had on Britain's rural way of life.

It's perhaps easy to understand therefore the creative impulses that deserted villages can inspire, from political discourse to more visceral evocations. There is a morbid allure inherent in the subject matter, a grim fascination that demands attention because of the sheer number of abandoned settlements that are present in our landscape.

It's estimated that there are over 3,000 deserted medieval villages in Britain, chronicling countless episodes of displacement and upheaval, some more traumatic than others. From 1069 to 1071, William the Conqueror's army laid waste to a huge area of northern Britain in what the landscape archaeologist Richard Muir describes as 'the most nightmarish and inhuman event in the history of the English people'. The genocide that took place in the 'Harrying of the North' is thought to have claimed over 100,000 lives and resulted in the systematic destruction of hundreds of villages. The Black Death that followed 200 years later had an even more devastating effect on the population. It is thought that over a third of Britain succumbed to the plague when it reached its zenith in 1348–50, with some regions experiencing an almost 90 per cent fatality rate. Despite the relative isolation of its villages, parts of Scotland were brutally affected, being hit by successive waves of the 'foul death' throughout the fourteenth and fifteenth centuries. Villages and hamlets whose existence was marginal at best were then swiftly abandoned by the survivors, often never to be recolonised again.

Likewise, many communities exposed to the vicissitudes of natural forces have often been comprehensively overcome. Over 4,000 years ago, on the Orkney Islands, storm-blown sands from the Bay of Skaill consumed the Neolithic settlement of Skara Brae, preserving it perfectly until 1850, when it was unearthed by another huge sea-storm. The place is beguiling in its halted presence and has inspired countless artists and poets since its discovery.

Life can be intimately imagined there. Passageways, once entirely covered, bend and wind between the eight houses and give the feeling of a burrow, a creaturely existence tucked away below the harsh asperities of the island's northern climate. Stone furniture still remains in all the buildings: beds, cupboards, dressers and fireplaces configured to family life. Warmth and noise and light still feel perceptible in each of the empty rooms.

The Sands of Forvie in Aberdeenshire retain similar secrets. In 1413, during huge gales, an entire village was engulfed by shifting dunes. Today only the chest-high walls of its church – the highest part of township – are visible. Such a biblical overwhelming left a predictable echo in local folklore. The obliteration of the village was said to be the result of a curse, invoked by the daughter of a former laird who had sought vengeance after being cheated out of her inheritance.

In this way, reality has often ceded to myth as a way of explaining the loss of villages and communities. Storytelling acts as both historical record and social catharsis, a means of both grieving and remembering. It is no coincidence, then, that Scotland – a landscape which carries the physical and psychological trauma of the Clearances – should be the setting for one of the most mawkish stories about a disappearing village.

The Broadway musical *Brigadoon* first opened in 1947 and was later adapted as an equally syrupy film in 1954.

The plot follows two American tourists on holiday in the Highlands who, while out hunting, stumble across a remote community not mentioned on the map. The pair are warmly welcomed and become immediately smitten with the curious old-world charm of the place. Only later do they discover the village's secret: that it is enchanted and appears just once every hundred years. On one level, the musical, the film and its countless re-workings since offer little more than an entirely caricatured, schmaltzy interpretation of Highland culture. Yet there is also something poignant about the choice of subject and setting. In describing a village trapped in time, a people and a place lost to the outside world but forever preserved in the moment of their past existence, we can register an undercurrent of real nostalgic longing. It offers something revealing about our relationship to the abundant histories of lost or deserted villages in Scotland, a collective subconscious pining for a place and a culture that no longer exists.

The deserted township of Bourblaige lay a couple of kilometres west of the bay. I cut uphill through a threadbare wood, the last of its coppery leaves trembling in the onshore breeze. On open ground I watched the sea heaving below me, swell patterns like gentle respirations, lifting and pulling at the water. Further out, the patterns turned to texture, catching the last of the day's sunlight with the tarnished brightness of brushed steel. I moved steadily upwards over sludgy tracks filled with hoof prints and bouldery ground covered in soft rushes.

My route was indirect, weaving between the volcanic lumpiness of the landscape, knolls the size of bungalows, faster to walk round than over. The topography crowded my view, denying a clear line of sight to the abandoned village until I was virtually above it. I turned the corner of a long, thin hummock, and all of a sudden it was there.

Perhaps it was the abruptness of my viewpoint, or the immediacy of the scene, but the sight of so many ruined buildings scattered across the hillside stunned me in a way I had not expected. I felt shock. But at that initial moment I was unable to properly understand why.

The clachan was spread over a wide area of flat ground, nestled underneath the south-western slopes of Ben Hiant overlooking the shore 100 metres further below. The structures were so numerous that it was impossible to total them all. I lost count each time I tried; twenty-nine, thirty, maybe more. They were clustered in small familial groups, a larger building with other, smaller ones nearby. I dropped down from the knoll and began to move among them. Low walls, some over a metre thick, rose barely above the height of the deer grass, so they too looked as though they were also sprouting from the moorland. Almost all had retained four walls and formed different-sized rectangles placed in varying orientations according to the flatness of the land. The largest were barely the dimensions of a living room, single spaces that would have housed an entire family; other smaller structures served as byres or storerooms.

The community had been set in the triangular area between two burns which converged at the southern end of the settlement. Each building was only a few strides from fresh water and just metres from the next house. I sensed something of the subtle agreements of space, the workable placement of every dwelling, the need for separateness, but also a desire for physical closeness, for human proximity and neighbourliness. It was, I realised, why the place had jolted me at first sight. I had not anticipated these unspoken arrangements still to be so vivid, so visceral. The landscape felt somehow paused because of it, filled with absence.

Bourblaige had been cleared in 1828, one of numerous mass evictions on the peninsula instigated at the behest of the landowner, Sir James Riddell. Riddell, like the succession of aristocratic landowners before him, quickly came to realise that the vast Ardnamurchan estate was an asset of limited financial reward – real estate which required significant management to extract any meaningful return. The advent of large-scale sheep farming, however, provided him the opportunity to drastically improve the commercial possibilities of his property.

By forcibly displacing the land's farming tenants in favour of sheep, far more income could be generated for much less effort. The decision (as was the case across so many Highland estates) was driven purely by motivations of economic self-interest but, crucially, had been enabled by a monumental shift in land rights and the rapid evisceration of Gaelic culture following the Jacobite defeat at Culloden.

Before then, pre-Clearance societies had largely operated on a patriarchal system of communal land-use deeply rooted in the relationships and mutual obligations between a clan chief and his kin. But with their status and power waning in the new, post-Culloden political order, clan chiefs desperately sought to maintain their standing in society, attempting, as the writer Madeleine Bunting observed, to 'navigate their way from paramilitary power to economic wealth'. They did so through the wholesale commoditisation of their ancestral lands, selling or leasing their estates, and transforming their roles from warlords and protectors to zealous landlords. Thus the age-old link between clan lands and the reciprocal duties of the chief and his dependents was catastrophically and irreversibly impacted. Up until that point, the belief 'that a piece of land could be owned in the same way as a horse or pot was not just inconceivable', commented Bunting, 'it was a denial of the ties of relationship which bound people together'.

And so, amid the countless individual human tragedies of the Clearances a deeper loss was also felt, a severing in the cultural and spiritual bonds between people and place. *Dúthchas*, the ancient Gaelic concept of attachment to land and the collective right to belong to it became an impossible idea, an expression that would find no modern translation in either language or society.

Riddell eventually acted on his commercial instincts. The first phase of Clearances began with the ransacking of the Camas nan Geall, Tornamona, Bourblaige, Skinnid and Choiremhuilinn clachans. Accounts tell of the landlord's factors carrying out their duties with workmanlike efficiency and a callous indifference to the local people. Roofs were taken off buildings, livestock and dogs were killed, and lazy beds ploughed up. The residents were encouraged to emigrate, or were moved to coastal settlements and placed on crofting land that was deliberately too small to sustain them, ensuring that additional work such as kelp harvesting would need to be undertaken for the landlord.

Little resistance was mounted to the changes. The Gaelic populations of Ardnamurchan found themselves facing into socio-economic and political forces far greater than any fight they could ever hope to muster. The relentless march of nineteenth-century industrialism, rampant capitalism and expansionist British imperialism was simply too much for a disparate feudal culture to countenance challenging. Any legal or notional claim to the land had also gone, evaporating a hundred years before when the estate of Ardnamurchan and Sunart had been sold by Clan Campbell. From then, the land had passed through numerous hands, transacted or inherited, until it had reached James Riddell, his grand plans, and his mounting debts.

A herring gull wheeled overhead, pitching its wings into the wind so that it glided its way towards the shore. I watched as it shaped its course, silently surveying the land below it: the empty sickle-shaped bay, the strange volcanic forms of the hillside, the ruins of the township half-hidden in the moorland, and the solitary human figure moving among them.

Islands of Industry

The Slate Isles, Firth of Lorn, Inner Hebrides

'Twice a day, the tide floods in like a wave.' He raised his palms and pushed his hands in a sideways motion across the Admiralty chart pinned on the wall of his sun room. 'It flows in from the Atlantic, reaches Ireland and then spills either side, pushing south round the boot of Cornwall and north over Cape Wrath and through the Pentland Firth until it meets again somewhere off the coast of Norfolk.'

'And in between?' I asked, pointing to the spray of islands breaking off like blast fragments from west of the Scottish mainland. He paused for a moment, stroking at his grey beard with thumb and forefinger. 'Ah, now that's where it gets interesting.'

He leaned closer into the chart and grinned. 'You see, the water here won't behave how you would expect it to. It doesn't simply just flow in and out with the tide. It swirls and jostles, and can switch direction according to variations in tidal strengths and the obstacles in its way. It's some of the most confusing and dangerous sea you'll find around the British Isles.'

It was for this reason that I had contacted Tony. He was a sea kayak guide who, in his words, had paddled the waters of the Firth of Lorn for longer than he could remember. Because of this, Tony knew the area intimately, its islands, its channels, its eddies and its ferocious tidal races. From years of observation, he knew exactly how the water could move at certain points and in certain conditions.

'We need to be careful here.' He drew his finger across the chart, over a slim flex of blue sandwiched between the southern point of Seil Island and the northern tip of the Isle of Luing, 'the Cuan Sound'. 'On an ebb tide, the water is squeezed through this narrow gap at seven or eight knots, and with wind against tide it can get pretty lumpy.'

I sensed 'lumpy' was perhaps an understatement, but the map was deceptive. Flat shapes of colour gave the impression of solidity and calmness: mustard yellow for land, cobalt blue for shallow water, sky blue for greater depth, and eventually white for the deepest trenches. It spoke nothing of the black-edged coastlines, the marble green of a churning sea and the creamy flotsam of breaking waves that we were likely to see on our journey.

Our route would take us four kilometres offshore to the furthest outlier in a small archipelago known as the Slate Isles. We would leave from Seil, an island separated from the mainland by the thinnest of tidal creeks, but joined to it by a single-arched, eighteenth-century stone bridge. From there we would cross the short stretch of water to the smaller neighbouring island Easdale and then push out south and west into the Sound of Luing to reach the tiny reefed-edged island of Belnahua. In this way, our journey would take us to ever-decreasing and remote landmasses. Islands like Russian dolls, each one smaller and further out than the one before.

The Slate Isles take their collective name from their geological renown, situated on a slender strata of metamorphosed sediment running north-east to south-west across the seaboard of the West Highlands. Fine-grained slate rock is found in abundance here in several rich deposits on Luing and Seil, as well as further north in the village of Ballachulish at the mouth of Glen Coe. Slate has been mined here since prehistoric times, and by the sixteenth century the area was held in such importance it was mentioned (as skalzie) by Donald Monro, High Dean of the Isles, who travelled through the Hebrides in 1549, noting 'ther a litle iyle, callit in Erische Leid Ellan Sklaitt', referring to what historians believe is modern-day Easdale, 'quherein ther is abundance of skalzie to be win'.

In the centuries that followed, small-scale mining evolved into massive commercial harvesting of the mineral resource. In 1772 Thomas Pennant recorded that slates of 'merchantable sizes' were being exported in quantities of 'about two million and a half . . . annually to England, Norway, Canada and the West Indies'. By the middle of the nineteenth century this number had grown as much as seven times, and the Slate Isles found themselves at the centre of a global industry, providing jobs to hundreds of workers and supplying roofing materials for buildings all around the world.

All that, however, would change. A series of natural and economic calamities, beginning with a storm surge of epic proportions, rendered the mines unworkable and the area destitute. Large swathes of the population left, never to return, and the land assumed a look of devastation, houses abandoned and machinery left to rust. It became another Scottish landscape defined by absence, the remains and relics the only indication of what had once existed there. Nowhere was this sense of scourge more evident than on the island of Belnahua. I had heard that a ghost village existed there; that rows of empty workers' cottages still looked out over the slate quarries which had flooded to form deep, azure-blue lagoons.

The thought of it was compelling. Not least because, following my visit to Bourblaige, I wondered if other such landscapes could be as evocative or affecting. Tony had been to Belnahua many times before, landing with clients to rest and to stretch legs. He admitted, though, that he had never explored the place properly, and was keen to do so. So, we agreed to spend time on the island: travelling with the tides and camping overnight.

On the day we met, however, it was clear that our plans would have to change. Thirty-mile-an-hour winds had blown in from the west, turning the sea into a welter of spume and spray – cresting waves and deep chop that were beyond my seamanship.

Instead, I headed out alone to the village of Ellenabeich, itself once also an island, hollowed out by slate mining until only its outer rim remained and now bound to Seil by the spoil from quarrying. There I would wait, hoping for a predicted drop in winds and a calming of the sea.

Ellenabeich translates literally as island of the birches. It's a name which has lost relevancy on both of its descriptive terms, for the trees are gone and the place can now be easily reached by road on an unbroken journey from the mainland. It refers now instead to a small village: three rows of whitewashed cottages sitting at the edge of a curving, slate-built harbour.

I arrived as the wind was gathering strength. Waves clipped against the shoreline, breaking in rapid succession. The air was moist and being pushed fast, slanting against me in a sideways drizzle. I saw only a handful of people as I wandered about, heads lowered and hoods up, scuttling between buildings. The place felt temporarily closed, battened down.

Signs of the mining industry were everywhere. A crane – iron cogs painted glossy black, its boom lifting at an angle above the houses – had been placed decoratively in the centre of the village bordered by raised beds full of brightly coloured flowers. Out in the channel between the village and Easdale Island was a march of wooden piles, the remains of a large pier once used for loading slate onto cargo ships. Above the harbour loomed the outcrop of Dun Mòr, its seaward face worn away to expose a dark scrape of glittering slate.

The same rock was also visible on the small beach, diagonal intrusions rising like black shark fins. Shattered pieces lay about them, rounded by waves into smooth oval discs that felt heavy in the hand. It was because of this that hundreds of people flocked here annually each September. The combination of an abundance of perfectly weighted flat stones and an expanse of calm water in Easdale's flooded quarries makes it the ideal venue to host the world stone skimming championships.

From inside a small stone waiting room on the harbour wall I pressed a button below a sign instructing how to call for the Easdale ferry. I wasn't sure exactly what to expect, but nothing happened. I waited a while and tried again. Still nothing. Five minutes later I pressed it a third time and lingered outside, this time noticing that the button had activated a light on the exterior wall which flashed in the direction of Easdale harbour for thirty seconds then simply stopped.

'Ferry only comes every half an hour,' said a man carrying shopping bags and wearing an oily sailing jacket. 'No point pressing that button.' I smiled and nodded my thanks, trying not to feel too much like a daft tourist.

When the boat arrived, I helped the man and his wife load their shopping on board and sat at the transom-end of the small ferry, the couple's border collie jumping on my lap as we bounced across the small channel. Near the stone jetty on other side, a collection of bright plastic wheelbarrows, fifteen or more, lay turned over in a bank of long grass, their tyres and resting legs pointing upwards, looking like colourful upended beetles. Some had numbers painted on the side, others had the names of their owners or the cottage they belonged to.

There are no cars on Easdale, partly because there is no need, but mainly because none could be accommodated anyway. From end to end, the island is small enough to walk across in ten minutes and is tightly packed with cottages, criss-crossed by narrow alleyways and grassy lanes. Residents go everywhere on foot, and the wheelbarrows, I guessed, were the main form of haulage, used for carting supplies up from the harbour and across the island.

I strolled up from the jetty, passing white buildings edged with drystone walls which held small wooden gates. There were picnic tables opposite, surrounded by pots full of flowers, a wooden-framed swing and a climbing frame in the shape of a boat. The track led up to a large quadrangle of open ground, bordered on three sides by terraced cottages. In the middle were two half-size football goals, white paint chipped back to leave rusted scabs, their nets sagging and breached with holes. At the far corner of the square, the grass had given way to a sprawling bramble patch. Next to it was a single, ancient lamp-post: ornate metalwork and a light kept within four glass window panes. It felt like a Narnian marker point on my entry to the island.

I wandered among the buildings noticing something of the life of the island. Individuality mixed with collectivism, shared spaces offset with the need for self-expression. Despite many of the cottages being identical, their doors and window frames had been painted different colours. Inside life also flowed out. Children's toys had been left where they were last played with and artworks painted on slates were propped against some of the buildings. I saw wind chimes and fairy lights hanging from the underside of roofs and bird feeders positioned in view of windows.

Functional items were all around too – wetsuits drying on walls, lawnmowers, wheelbarrows, and a fishing net strung up on poles with lobster creels stacked around it. Everywhere I went there was a mingling of the communal and the contained, the quaint and the quirky. At one point I came across a signpost with direction arrows on each of its four sides. I approached it, hoping it might provide some sense of orientation to the village's compact layout. The destinations, however, turned out to be much longer ranged: Paris 715 miles; Moscow 1,746 miles; Lima 6,828 miles; and Sydney 11,275 miles. There was one exception, a sign halfway down the post: 'Pub', the pointer indicated, '48 feet'.

For almost an hour I pottered about, mooching along gravel-pathed vennels that had become overgrown with rose hips and the fiery orange-flowered montbretia. At times it was hard to know where a lane ended and someone's garden began. Unbordered lawns appeared suddenly, and I stepped quickly in front of houses, conscious not to invade privacy, but also becoming familiar with the loose notion of property and ownership on the island.

Back at the sea edge, above one of the island's flooded quarries, I tramped down a space to sit among some long grass. It was a calm spot, below the worst of the wind. The smell of woodsmoke pulsed in on gusts from the village, and from behind me I could hear the soft roar of wave on rock. The quarry's water below was still. Its depth had rendered it so dark that it threw back the world above it in silhouette: a wood-block print, black-and-white patterns of sky and bank.

The Slate Islands are defined by their mining history. The industry first determined the archipelago's titular relevance, but, perhaps more pertinently, permanently transformed the islands' topographical appearance. Land has been created here: causeways, embankments and moles formed from slate spoil edging long fingers out into the sea to be shaped and moved in the tides. And land has been taken away: gauged and scooped, voids created, quarries in the landscape like hole-punched paper. The islands' present-day manifestation is therefore a product not only of their geological circumstance but also of human intervention.

The earliest mining extractions were made on the islands' foreshore. At low tide, hardwood wedges were inserted in the natural fissures of the exposed rock. As the sea rolled back with each rising tide, the stakes would swell repeatedly in the water, cleaving gaps in the slate as they expanded. The technique was crude, but effective. With the openings that had been formed, combined with the strength of the rock, large flat sections of slate could be easily broken out, ready to be split and trimmed into smaller pieces. The individual slates were then sized according to the various dimensions required for roofing, a 'princess' being the largest (600 × 350mm), followed by lower ranks of nobility – a 'duchess' (600 × 300mm) and a 'countess' (500 × 250mm).

As the mining activities increased, large parts of the islands began to fall below the level of the sea. At the shoreline, walls and sluices were built to control the flow of tidal water, allowing the miners a few precious hours of dry conditions. The job was hard, with teams of six men working continuously in the damp and the cold: two quarriers mining the slate at the rock face, a couple of labourers to transport it above, and a pair of nappers whose job it was to cut the slate into the finished tiles. Women and children were also involved, carrying heavy loads of slate to the harbour in nothing more than creel baskets strapped to their backs.

By the middle of the eighteenth century, demand for the islands' slates was surging. The rapidly expanding cities of the Empire required roofing materials in unprecedented quantities, and ways of upscaling production needed to be found. The initial, rudimentary practices of extraction quickly became industrialised, with the addition of machinery intensifying production. Steam engines were used to pump the quarries dry, making it possible to work the mines at increasingly greater depths below sea level; platforms, cranes and winches were built to hoist the slate to ground level; and explosives were introduced to break out sections of the rock. Tram lines were also later laid, with track systems linking the quarries and the harbour.

Human labour, though, remained the main productive

element necessary for the mining, and a large, skilled workforce was settled on the islands. From 1745 onwards, cottages for the quarrymen and their families were constructed on Seil, Luing and Easdale. The accommodation was basic – squat buildings, with thatched roofs stooped low in the landscape – but of a decent standard, usually consisting of two rooms with additional space for sleeping in the roof. Alongside them, communities soon developed. Churches and school buildings were established in each of the main quarrying localities, with a medical officer employed by the mining company to oversee the health of the workers.

Despite their geographical remoteness, most of the islands grew to support a thriving population, regularly connected to mainland Scotland and the rest of the world by the huge volume of shipping required for exporting the slate. For a while, the Sound of Luing became one of the busiest waterways in Britain. Local archivist and writer Mary Withall has recorded that in 1825 the ships using Easdale harbour alone 'amounted to seven brigs, 15 schooners, five galliots, 254 sloops and 245 steamboats'. The inflow and outflow of human traffic was also significant, with itinerant workers, residents and even tourists regularly arriving by passenger steamers that moored on the wooden pier in Easdale Sound.

For all this activity, though, the residents of the most isolated of the islands, Belnahua, still maintained a marginal existence. Living on a section of land that was half the size of Easdale and which was situated much further out to sea, the island's community depended on supplies of food and fresh water being shipped across from Seil and Luing. Reaching the island, however, was often a precarious undertaking. Fast-running tidal channels and a lack of natural or man-made harbourage meant landing on Belnahua was only possible when tides and weather permitted, and then only by smaller craft run ashore on its narrow, shingle beaches.

In such a borderline territory, survival became a delicate balancing act, easily imperilled when the wild forces of the Atlantic gathered strength. Such a moment arrived in the winter of 1881, when the Slate Islands were hit by a sea-storm of epic proportions. Accounts at the time render the episode with tones of Biblical calamity. Easdale felt the full wrath of the storm, and for several hours became submerged by the rising tide. 'The sea swept over the island about four o'clock on Tuesday morning and caused great alarm to the inhabitants (some 400),' reported the *Oban Times*. 'The cry was raised to make for the hills but this was found to be impossible from the severity of the storm, and the rising of an unprecedented high tide.' The situation was dire. Unable to reach higher ground for fear of being swept away, and with no hope of launching their boats, the islanders had little option but to stay in their swamped homes, 'piercing holes through the roofs of their dwellings, there to storm it out to the last'.

Those hours of darkness trapped within the maelstrom must have seemed the longest of the islanders' lives. But with daybreak came salvation; a turning of the tide and a receding of the worst of the flood waters. The full devastation of the night's fury then also became apparent. Remarkably, no lives were lost, but the storm had otherwise been utterly ruinous. 'The slate quarries, of great depth and girth, were brimful of water immersing the Lessee's plant and workmen's tools.' In addition, the 'landing pier was swept away together with a great quantity of cut slates lying ready for shipment'.

The necessities required for self-sufficient island life had also been obliterated. Outbuildings were destroyed, their contents, including cattle and pigs, washed away, many to be later found 'lying high up on the bank some miles off'. Small boats, so vital to the island, were similarly lost: forty in total, smashed to pieces by the waves or driven out to sea.

The impact of the great storm on the slate industry was catastrophic but did not deliver a fatal blow. Ellenabeich quarry had been completely flooded, with much of its equipment lying irretrievable, over 200 feet underwater. The mine would never be worked again, but machinery that could be salvaged was transferred to Easdale's quarries, and as operations eventually returned to normality, the islands' slate exports recovered. For a while at least, the Slate Islands' industry continued to survive.

The reprieve, however, would not last. Forces were at work that would ultimately be far more damaging than even that of the great storm. Competition increased from the Ballachulish and Welsh quarries as well as from the importing of cheaper roofing materials from abroad. Architectural tastes were changing too, with clay tiles being used more often and replacing slate. Faced with the cold disregard of free-market dynamics, the Slate Islands struggled to find a response.

Production, which was always so dependent on the available workforce, was gradually debilitated by the slow exodus of workers leaving the Highlands and Islands. The loss of the male, working-age population was continual and absolute, with men tempted by employment in Scotland's cities, the merchant navy and even the lure of finding fortune in the gold mines of America, Africa and Australia. Military service was just as devastating, with men pressed in to service first in the Boer War and then, with more finality, the Great War. With only old men, women and children left behind, the islands' slate industry soon reached the point of no return. Life in the remote communities – for decades only fragilely sustained by mining – also became an impossibility.

It was late afternoon, and with time to kill before the last ferry off the island, I made my way to Easdale's only vantage point: a hill, barely a hundred feet high, but shaped with crags and gullies and rising in an elegant ridge curving all the way to its summit. I walked the edge of the island, passing more flooded quarries – atolls of calm water hemmed in by thin reefs of tide-marked rock. One pool was beautiful. Its circumference glowed brightly with an aquamarine hue. It reminded me of An Lochan Uaine, the small loch hidden in the Glenmore Forest that pulses jade green through the trunks of ancient Caledonian pines.

From the top of the hill I sat and watched a RIB powered by two large outboard motors track a course round the island, bouncing over waves and kicking up spray with each thud of its bow. The pilot dropped the engines to a low hum and held the boat facing into the oncoming swell. Through binoculars I could see two other people. A man leant overboard reaching for a buoy, while a woman in waterproof overalls clung on to his waist, anchoring him to the deck. They were retrieving lobster pots, pulling on wrack-covered lines to check their catch.

Further out, I scanned the outline of Insh Island, a long, rocky outcrop three kilometres north of Easdale. I had heard about this place several times before my trip. It was, I had been told, home to an occasional hermit. Some stories had referred to him as the eccentric owner of the island who made his home in a sea cave on its eastern edge. Other versions described him as a wealthy banker, arriving from the City every year and being shipped across with supplies to stay on the island for the summer months. Most accounts portrayed him shouting angrily at boats that tried to come ashore, but every retelling was consistent in the fact that he always appeared stark naked.

That night I wild-camped on the domed outcrop that loomed above Ellenabeich. I picked my way up a steep gully of loose rock, then onto grassy flanks nibbled short by sheep, pitching my tent on a promontory of flat ground that looked out across the Firth of Lorn. A ewe and her lamb stared at me vacantly as I fussed around my campsite, then continued to watch as I sat and ate dinner from my mess tin.

The day was darkening, grading towards night. Out at sea, the water was grey and foam-flecked. I could see the islands of the Firth: Fladda with its lighthouse; Lunga, barely visible in front of the larger mass of Scarba; and the Garvellachs, hazy on the horizon. In the foreground was Belnahua. Its profile was long and flat with a knoll of higher ground two-thirds along so that from a distance it gave the uncanny resemblance of a submarine surfacing. I visualised my route to the island, plotting a course out from Easdale where, one by one, the lights of the village had started appearing in the dusk.

At some point in the night the wind must have stopped, for I fell into a deep sleep undisturbed by the ripple and hum of the tent's flysheet. I woke late, roused by the sound of sheep and the heat of sunlight filling the tent. Outside, clouds of midges had formed in the still air; the sea beyond was calm, its surface glossy and teal-coloured.

An hour later I met Tony by the harbour. He wore an old baseball cap and sunglasses and was smiling broadly. 'Glad to see my weather order came through,' he said lifting his arms and looking up at the blue sky.

Tony talked excitedly about tides and weather as we unloaded our equipment into the small, pebbled bay below the pier. The village around us was busy. People milled about as we kitted out the kayaks and watched as we donned our drysuits. The water too was full of activity. We timed our exit from the harbour carefully, avoiding the incoming ferry and two small dinghies that were motoring across Easdale Sound, paddling alertly until we reached the nearest of the Easdale's flooded quarries. It was high tide, and we were able to scull in over a low wall of rock that normally sat above sea level.

The place held a deep-water calmness, sound and light absorbed and condensed within the steep-sided pool. It was because of their huge volumes of vertical water and their exceptional clarity that the sea-filled mines were now often used as training venues by divers. Ellenabeich's quarry was particularly deep and, according to Tony, the machinery lost in the 1881 storm could still be seen resting at the bottom.

We left Easdale's shore and began paddling south-west into open water. To our left was the wide sweep of coastline: the western edge of Seil Island, then further south,

Luing. I could make out the white frontage of houses at the land's edge, the brindled colours of moor and cliff behind them. Jellyfish floated past in our wake, tremulous blooms of purple and white. Then, a kilometre away, a porpoise surfacing, its body curved in a perfect arch as it leapt from the water.

Ahead of me, Tony pulled up alongside a small, orange marker buoy. 'Do you see what's happening here?' I looked blankly at the bright plastic sphere, unsure of what I should be noticing. He pointed to water next to the buoy. A V-shaped eddy, barely discernible, had formed on the surface. 'Five minutes ago in the harbour, the sea was moving in the opposite direction; now it's flowing the other way. The turning of the tide here happens in an instant.'

We travelled diagonally across the current, paddling for several kilometres using a series of buoys to plot our course until we reached the largest, a tall Admiralty marker that warned of a reef to our west. Belnahua was close now, little more than two kilometres away, a thin slice of land that sank out of sight with each rise of the soft swell. Next to it was another island, Fladda. A low scrape of rock impossible to register without its squat lighthouse.

For an hour during the night I had watched the lighthouse's beam from my crag-top camp, fixated by its steady rhythm. Now though, in daylight and with a seal's-eye view, the building commanded my attention in a different way. It appeared strangely unfounded, as if constructed on nothing more substantial than the air hazing above the water – an act of levitation enabled by the deadly rocks it was built to warn against.

Glen Loin Caves, Succoth: refuge of early climbers
(All photographs courtesy of the author unless otherwise stated)

The navvies' graveyard at the Blackwater Dam

The dam-edged western end of the Blackwater Reservoir

Jock's Road descending from higher ground into Glen Doll

The tiny Davy's Bourach shelter on the Jock's Road drovers' route

The canoe exploration of Loch Awe's islands: on the small beach of Inishail island

The cemetery on Inishail, one of many burial islands in Scotland

The remains of the replica of the Atlantic Wall, Sheriffmuir

An illicit still hidden in Ardnamurchan

The Clearance village of Bourblaige

Paddling across the Firth of Lorn towards the island of Belnahua, Inner Hebrides

The flooded quarry on the abandoned island of Belnahua

Military ruins on Inchkeith Island, Firth of Forth

The unique geology and landscape of Assynt

The Bone Caves, Assynt, and the Allt nan Uamh stream

The Clootie Well, Munlochy (Geopx/Alamy)

By the end of the eighteenth century, Britain was struggling with a problem it neither fully acknowledged nor had an answer to. Throughout that century, maritime trade and naval traffic had increased at an incredible rate. Sea routes around every part of the British Isles were being used more than they had ever been before. Journeys to and from the New World were providing the country with a vital supply of commodities as well as routes to ever-more lucrative new export markets. Business with Europe was booming, with a regular stream of shipping leaving Britain's shores for Scandinavia and the Baltics, as well as France and Spain (when hostilities permitted). Meanwhile, the growth of industrial Britain meant a huge increase in raw materials arriving by sea and finished products for export leaving the same way. Elsewhere, large fleets of ships serviced the whaling industry, while countless numbers of vessels were in the employ of the slave trade. Against this bustling commercial backdrop, the Royal Navy was also a constant presence, embroiled in long-running battles with foreign nations for control of the seas surrounding Britain.

With vast numbers of ships plying their respective trades off the coast of Britain, but with few accurate maps to chart the often treacherous waters, shipwrecks were a horrifyingly common occurrence. In the 1790s an average of 550 ships were lost off British shores each year. The number was to rise steadily in the nineteenth century, and by the 1830s it was estimated that on average at least two ships were lost every day.

Despite the huge amount of deaths that were occurring at sea, the hardships and mortal peril that sailors faced were largely overlooked. There was a grim acceptance of the danger that came with life at sea, both from the mariners on board and the merchants who owned the ships, each knowing that it was mostly a matter of chance if their lives or their cargo were to be lost. Likewise, the government had exercised a similar, long-standing indifference to the situation. It wasn't until the 1780s when public agitation and lobbying by shipowners reached a critical mass that an Act of Parliament established the Northern Lighthouse Board (NLB) in 1786.

Lighthouses had existed in England before that time, but it was Scotland's coast – over 4,000 nautical miles of storm-swept headlands, unpredictable tidal straights and unmarked islands – which proved a formidable and unassisted challenge to navigate safely. Such was the difficulty of finding safe inshore passage to Scotland's harbours that sailors required considerable skill and local knowledge to identify landmarks and potential dangers along a particular stretch of coast. Making landfall after dark was almost impossible, with few reliable beacons to guide towards even the largest of ports. The accident rates for the busiest waterways were incredible. In the Firth of Tay alone – a large estuarine entry-point notorious for its rocks – seventy ships were lost over the course of 1799.

Something had to be done, and the NLB turned to Thomas Smith, a lamp-maker from Edinburgh, to build the first of their lighthouses. Smith was an unusual choice. Although he had proven himself as a notable innovator, designing a powerful parabolic reflector light to improve navigation in the Firth of Forth, Smith had none of the engineering training or experience necessary to plan and execute the construction of buildings in remote locations. Like the descendants that were to follow him, Smith was a man of deep determination and immense practical resolve.

He immediately set about filling the gap in his engineering knowledge by undertaking a crash course in lighthouse construction with the English lighthouse-builder Ezekiel Walker, and over the following decades would develop and refine his own expertise through a series of commissions throughout Scotland. There was a pioneering spirit to this early period of Scottish lighthouse-building. Lessons were learnt from one project and applied to the next. It was a proficiency that Smith not only accumulated for himself but also shared with his stepson and apprentice, Robert Stevenson, thus beginning an engineering dynasty that would last another four generations.

The Lighthouse Stevensons, as they came to be known, would eventually be responsible for the construction of all of Scotland's lighthouses. For more than a century, as the position of Engineer to the NLB was passed down through members of the family, each of the Stevensons would, in their own way, push the boundaries of invention and challenge the realms of engineering possibility. Robert's greatest achievement came relatively early on in his career.

The construction of the Bell Rock lighthouse was an engineering marvel, audacious in its sheer improbability. Built on the Inchcape Rock, nineteen kilometres off the coast of Arbroath, the reef was fully submerged at high tide and only barely exposed at low water. Work was dangerous and time-bound. Heavy seas would regularly render any attempt at landing on the rock futile, while access would cease twice a day as the tide came in – the workers having to retreat to their living quarters on board a ship moored a mile away.

On one particularly terrifying occasion, the ship broke free of its anchor and began drifting away while all thirty-two men, including Stevenson, were working on the reef. With only two small tenders tethered to the reef, most of the men would have drowned as the rising tide engulfed the rock, were it not for a supply boat which by luck arrived just in time.

After three years, the project was completed to widespread public acclaim. Sir Walter Scott visited the lighthouse, claiming 'no description can give the idea of this slight, solitary, round tower, trembling amid the billows'. Turner would also later paint it, the image of the structure standing resolute and steadfast amid waves that break as high as the light itself.

Robert's reputation was secured. He continued building lighthouses, and although none could ever match the Bell Rock, his sons would each create their own masterpiece to rival the technical genius of their father. Alan Stevenson was responsible for Skerryvore, a lighthouse built on a jagged reef in uncharted waters nineteen kilometres off the island of Tiree, while his brothers David and Thomas erected Muckle Flugga, Britain's most northerly lighthouse perched on a sharp protrusion of rocks in the Shetland Isles, as well as isolated stations far out at sea on Chicken Rock and Dubh Artach.

There remains something profoundly tenacious about the Stevensons' achievement – a stubborn overmastering of the ferocious boundaries between Scotland's land and sea: the creation of outposts, light-bringing strongholds that tell of lives saved and lives that have been lived at our country's wildest and most remote peripheries.

The individual and personal histories of these places are as diverse and intriguing as the lighthouses themselves. Isolation and immersion in a remote and wild location proved to be both a compelling and a maddening environment for the keepers, and many accounts exist of deteriorating mental health for them and their families. Dhu Artach light in particular, situated on a skerry twenty-four kilometres south-west from the Ross of Mull, was so remote it was referred to by the keepers as the Black Hole. Its seclusion was enough to drive one of its residents to attempt to leap from the rock and swim back to the mainland.

Darker events were also recorded. In 1960 an attendant murdered his fellow keeper on Little Ross light in the Solway Firth, while in 1900 all three of the keepers at the Flannan Isle lighthouse inexplicably vanished. To this day the incident and circumstances of the disappearance remain a mystery. When the relief keeper landed on the island, he found it eerily deserted. Crockery and cutlery had been cleaned and left to dry in the kitchen; the hearth had been prepared for a fire; and the building's light was in perfect working order, freshly cleaned with a full supply of oil. Strangely, though, all the clocks had stopped, and the log books, which had otherwise been kept up to date, had received no entries for a week.

There were also numerous tales of heroism and courage, including the rescue of twelve seamen, whose ship was wrecked on the Pentland Skerries, as well as the snow-bound journey of the keeper from the Fair Isle North lighthouse who battled his way through storm-force gales to help the neighbouring Fair Isle South lighthouse stay lit following a bombing raid in 1942.

Life at the lighthouses was undoubtedly hard, but it also enabled a rarefied existence, detached from mainland timeframes and social structures, and intimately acquainted to wild surroundings. 'Could I live like this forever?'

pondered Peter Hill in his memoirs as a lighthouse keeper, with his view 'a golden hue to the cliff face, speckled with the white of nesting gannets and framed by the oh-so-blue sky'.

The lighthouse I could see at Fladda had the best of both worlds. Its position, built on a small island sitting off a larger island, was certainly isolated. There was no water supply, and the channel separating it from Luing, although short, could often be uncrossable. But the island maintained an unusual community, distinct from many of the NLB's more hermetic settlements: it was large enough to accommodate the lighthouse and several buildings, encircled by a vast sea wall, thirty feet high and fifteen feet thick in places. Two households made their home there. At one point, nineteen people lived on Fladda; men, women and children, all presumably able to coexist in relative harmony, despite being hemmed in by the prison-like walls and the wider entrapment of the sea.

The sense of the island as a commune was also present in stories I had heard of its garden. Inside its white ramparts was earth that was richer than anything else on the surrounding coast. It had been deposited there as part of a regular transaction, soil for slate. Tonnes of fertile loam used as ballast in Irish cargo ships was dumped on the island before the rock could then be uploaded for transportation. It meant that Fladda became famed locally for its produce – fruit and vegetables cultivated by its small lighthouse population that were unrivalled by anything grown onshore.

Fladda's island life, though, would permanently change in 1956, when automation of the lighthouse ended the requirement for human presence on the small skerry. It was a narrative that would play out across all of Scotland's lighthouses over the following four decades in a gradual abandonment of all of the Stevensons' wild eyries. On 31 March 1998 the last manned station, Fair Isle South, became fully automated, bringing to a close 211 years of life at the furthest edges of Scotland. It was the end, as the writer Bella Bathurst puts it, 'of our Lilliputian effort to pin down Gulliver . . . to civilise the sea, to make it manageable and rational'. The empty buildings now bear strange testament to all those years of encampment and the human struggle to make safe the sea for others. Instead they have become sentinel monuments, vacated engineering masterpieces, now just whirring to the click and hum of machinery.

The current pulled us round to the south side of Belnahua with such force we were able to drift without paddling. We found an eddy and a small bay with charcoal-grey sand where we could land, heaving the boats above the tideline. Higher up the beach's incline, shingle turned to pebbles, then to rounded plates of rock, the size of discuses. I turned one over in my hand, feeling its weight and noticing it glimmer. Gold squares studded the surface, embedded with all the precision of a master jeweller.

The flecks were iron pyrite, mineral inclusions formed during the rock's metamorphic processing. The stone sparkled as I tilted it, and when I angled my head to a different view of the beach, a similar effect occurred in multiple: the slates glinting with thousands of metallic reflections. In some rocks the lustre had faded, the pyrite rusting to leave chestnut-brown stains, or completely dissolving so that small holes were left in the slate. I brought one close to my face, holding the gap to my eye so that it framed the beach, the sea and the distant coastline.

We left the shore, crossing through a gap in a thick wall that stockaded the southern side of the island like a broch. The place was pathless and free of livestock. Grass had grown unchecked and was so long that it fell in lank piles, soft to the tread but tangling underfoot. Walking was hard work and required the same exaggerated strides of moving through drifted snow.

The deserted village was right in front of us, clusters of hollow-eyed buildings with window and door spaces showing light-filled interiors. We moved first to a line of terraced houses similar in appearance to the inhabited cottages I had seen on Easdale the day before: two-room structures, built from a haphazard mix of slates and red brick that looked ready to collapse at the slightest touch. The roofs had long since disappeared, and quiffs of tufted grass grew along the tops of the walls.

I stepped inside the nearest building, careful not to dislodge any of the masonry. The rooms were compact: barely twelve feet square, with just a single hearth at one end. Where the grass had not fully taken hold, I could see that slate had been placed on the ground; misshapen pieces tessellated as best they could to create a makeshift floor. In the gaps in between, ragwort grew in bright yellow constellations. Despite the cramped simplicity of the cottages, each had an incredibly open southerly aspect. Their windows faced out on to a small beach and, beyond that, to islands which overlapped and receded into the distance in softening blue tones. I imagined the intense contrast that came with living there. Only metres from the water and lashed by the worst of the winter storms, yet in softer summer moments, hours must have been spent gazing at that view.

Behind another terrace, at the edge of a shallow pit, we found machinery. An upright carcase of orange metal, propped up by its own weight and leaning at a slight angle. It had the look of a large, felled animal. Tony began an immediate examination. 'Gears, pistons, crankshaft,' he muttered to himself as he worked his way round its structure. 'There's an incredible amount of technology here.' Most of its shape was intact, but where the rust had taken hold most damagingly, it had corroded the metal in flakes of deep umber resembling charcoal or the preserved texture of timber pulled from a peat bog.

'A steam-driven crane,' Tony concluded after several minutes of scratching his beard and visualising the mechanism's movement. 'And there's the boiler,' he said, pointing to the rib cage of a large cylindrical drum lying at the bottom of the pit.

Close to the crane – because everything on the small island was only a stone's throw away – we clambered to the top of Belnahua's highest point, a prism-shaped knoll with a large cavity cleaved out from its southern end. The climb was tricky, the incline steep and covered in slippery heather roots. Small insects and tiny brown butterflies fizzed up from the purple flowers, clogging the air around our eyes. The summit was a miniature Alpine arête, a sharp spine of rock with precipitous drops on either side. At the apex of the ridge I found a safe position to sit and scanned the island's geography.

Directly below me was the slate mine, a single flooded quarry of dark water that filled the centre of the island like an ink well. The gouged area was so large that in places only a narrow boundary of land had been left to buffer against the sea. In the middle were two protrusions of rock, islands within an island: long, thin and tapering at each end with bright green grassy tops that resembled the backs of basking crocodiles.

From up high I could see the buildings we had passed as part of the larger settlement, established around the bottom half of the quarry. As well as the terraced cottages I had looked in, there was a scattering of other buildings: pump houses and engine rooms, coal sheds and outhouses. On the eastern side were two more substantial-looking buildings – a two-storey house and a bothy recently painted white and roofed in corrugated metal. Near where we had walked, I also noticed something on the ground, a rectangular outline only now visible from above. It was the earthworks of a large, deliberately shaped piece of land, ramped on four sides by a low embankment – a communal space for growing vegetables or maybe even for playing.

I had read somewhere that at the peak of its population, the island was home to over a hundred people, many of whom were children. I tried to imagine life in such a contained space, the claustrophobic freedom of a child being able to explore anywhere but simultaneously confined everywhere. For the adults too, it must have been a life adapted to compromise and togetherness, relationships both bonded and frustrated by the physical closeness.

To the north of the island I saw the extended finger of a reef gradually revealed in the receding tide. The rocks edged out into the Firth in two parallel lines as if they were once part of an ancient slipway. Even when they disappeared from sight, their presence was still marked for hundreds of metres out to sea, creamy streaks and swirls rucking the surface of the water – a submerged tendril reaching out from Belnahua, as lethal as it was inconspicuous.

On the night of 26 October 1936, a Latvian steamship journeying up the Firth of Lorn was wrecked here and sank almost immediately with the loss of nearly all its crew. The *Helena Faulbaums* had been travelling in ballast from Liverpool, bound for Blyth in Northumberland when she became caught in gales of exceptional severity. So extreme was the storm that villagers on nearby Luing later recounted being unable to open their doors as sea-spray rained down like 'gravel beating against the houses'.

With such a light vessel, movement through the high seas of the Firth had already proved near impossible, but disaster struck when the ship's steering mechanism failed. Unable to maintain any course, the *Helena Faulbaums* drifted by the beam until she ran aground off Belnahua, where the decision was taken to abandon ship. Fifteen men lost their lives, but remarkably four managed to survive, reaching the island's shores and finding shelter and old blankets in the deserted cottages of the slate mine.

We continued exploring, walking the rim of the quarry. I studied the visible details of the flooded mine, searching for some of the dozens of archaeological features that apparently still resided somewhere in its depths: tram tracks, slate bogeys, machinery, tools. With no flow or sediment, the water held an incredible clarity, magnifying and focusing objects below the surface into a realm of intense hyperreality. Shelves of slate became blue-tinged lagoons, the rocks within them, although feet below the surface, appearing close enough to touch. It was a dangerous deception, a rare optical lucidity that is recurrent in many accounts of drownings. As we crossed a scree run which sledged diagonally into the pool, I watched as a clattering stream of rocks broke the water, momentarily warping and distorting the sapphire-edged shapes below.

On the far side of the quarry at the island's eastern side, we found more machinery. This time a prone skeletal engine: limbs of interconnecting levers, rods and cylinders that seemed to push and pull at each other even when devoid of motion. Incredibly, the heavy-looking apparatus still rested on its original joists, now nothing more than brittle matchsticks of decayed timber. From the space below, saplings had taken root, white-barked branches rising up between the rusted metal. I left Tony poring over the mechanism and crossed back on to the shore, still trying to make sense of the place.

Even though a hundred years had passed since it had last been occupied, the island still harboured a feeling of rapid abandonment, of buildings suddenly vacated, machinery hurriedly forsaken, tools dropped from hands. I wondered at the islanders' acts of leaving, the urgency of those departures. Had there been time to reminisce, to pay a final visit to a favourite spot, to quietly say goodbye? Or was the beckoning of a new life too strong, too insistent to allow time to think about the past? There may have been an imperative, now unknowable; only a specific, thin moment of opportunity on a particular day to make good an escape across the fast-flowing tides.

Further along the beach I came across another discarded boiler, settled on a terrace of flat ground at the top of the shore's incline. A ramp of storm-washed pebbles had built up against its seaward side, delicately bracing it from rolling into the water. Lines of rivets ran around the outside, small half-globe nodules that were so perfectly ordered they looked almost decorative. At the top of the boiler's curve, rainwater must have collected gradually over time, for the thick carapace had rusted away leaving cartographic-shaped openings: elaborate gaps that resembled sea lochs, each with their own intricate metal coastlines.

The ends of the cylinder were missing and, from the nearest opening, a spray of plant life – campion, seathrift, grasses and dock leaves – tumbled out, as if part of some deliberate floral arrangement. A handful of slates had also been washed inside, scattered around the base in an uncanny accessorising of the display. In the boiler's arched ceiling, gossamer threads of spider webs were strung across like aerial fishing nets. I noticed something else. Clumps of grass had been packed in the centre, uprooted from elsewhere and now yellowed to straw. The stalks had been compressed and lay shaped in a soft circular mesh. A nest. The boiler had become a home to one of the island's seabirds, converted into a place of cover and safety. It felt like an unintended but happy rewilding, the repurposing of Belnahua's long-forgotten human past for a wilder present.

Tony and I met back at the boats. We sat and talked for a while, drinking coffee and sharing chocolate. He told me more about the islands of the Firth. About the 'Grey Dog', the tidal race between the islands of Lunga and Scarba, where the water runs at eight knots in full flood, and about the infamous Corryvreckan, the whirlpool north of the Isle of Jura that had once almost claimed the life of George Orwell. He also spoke of another uninhabited island, Eileach An Naoimh, a thin strip of cliff-stacked land where strange, corbelled beehive structures look out onto the Atlantic telling of wild hermitage and an ancient religious devotion.

Sea Fortress

Inchkeith, Firth of Forth

We were five kilometres offshore, and the sea was eerily calm. Beneath my kayak the water felt supple, gently flexing in fat swells, its surface unbroken and quartzy. It was mid morning, but the day was already exceptionally warm. Clouds as transparent as dandelion clocks hung immobile in the sky. Small flies hovered and whirled near my paddle blades, gyrating around each other in tiny airborne clusters. A cormorant crossed noiseless above us, its black wings sleek in the sunlight. Ahead of me, and beyond my companions, I could see the island's outline: a long hump of rock skimming the horizon, the edges hazy and moving in the heat, giving the impression it had somehow been set adrift.

It had taken a long time to make this journey. For the twenty years that I had lived in Edinburgh, the thought of it had always been there. A venture, permanently half intended that would appear in my peripheral vision again and again like the island itself, constantly visible but so familiar as to be ignored. Occasionally, though, I would catch glimpses of the island framed in ways that would stun me with sudden, unerring interest. The sight of it from high on Arthur's Seat, its remote shape, black and brooding in the Firth of Forth. Or bordered perfectly between a gap in the stands at Easter Road Stadium, an island vignette, seen curiously out of context at a football match.

I had also flown over it many times. On homeward-bound planes which banked low over the coast, I would get a gull's eye view of the island. It would give me the chance to survey its complex, post-apocalyptic townscape: deserted roads and decaying buildings, greenery threading through concrete, shrubs emerging from the rubble. It was a contradictory but compelling mix. A strange scene of urban dereliction in an elemental and unconnected place.

That the architecture and edifices of different times could still exist so forgotten and untouched, yet so geographically close to Scotland's capital, seemed paradoxical, but also made perfect sense. For centuries, Inchkeith's wild form and isolated position – a craggy, kilometre-long, hill-topped rock in the wide, storm-blasted waters of the Forth – had both alternatingly defined its importance and rendered it a landscape that could easily be cast aside. Over the centuries it had served a multitude of purposes. Fortress, garrison, farm, lazaretto, prison, religious settlement and even the venue for a bizarre linguistic experiment. As backstories went, it could certainly claim an eventful and troubled past, and more than conformed to the writer Patrick Barkham's contention that 'a few square miles of island loom far larger than the equivalent pocket of land on the mainland'.

The same was also true for the other islands scattered in the Forth, the 'emeralds chased in gold' of Walter Scott's epic poem *Marmion*. Each, I had found out, had their own rich narratives, outsized stories that were much bigger than their contained landmasses would suggest. They were tales of human and natural history – warfare and piracy, extinction and regeneration – that were intimately entwined with the lives of people far beyond their immediate watery domains. In this sense I had come to regard Inchkeith as something of a landing point, literally and figuratively, fixed amid the change and flux of wild histories I had overlooked for years.

On 14 September 1779 panic descended on the towns and villages bordering the Firth of Forth. A squadron of five ships under the orders of the infamous naval commander John Paul Jones had been spotted lurking at the mouth of the estuary just off the coast of Dunbar. Fearing imminent attack, defences were hastily assembled. The town's magistrate appealed for military assistance and a regiment of dragoons was swiftly despatched from Edinburgh. Meanwhile, local men rallied to mount an improvised resistance. On one of the hills overlooking the waterfront, embrasures were quickly built to house a battery of guns, while a twelve-pounder cannon and two other guns were placed at the entrance of the harbour.

The alarm caused by the sight of the ships was not without good reason. John Paul Jones, sailing under a commission for the Continental Navy of the American Colonies, had a fearsome reputation: he was well known for attacking British naval and merchant vessels as well as launching land-based offences.

Jones had been born in Scotland but had emigrated to America as a boy and later sided with colonialists in the Revolutionary War against the British. His early exploits on behalf of the newly formed navy sailing off the eastern seaboard of the Americas had proved highly successful and established his renown for nautical warfare. As well as raiding the port of Nassau in the Bahamas, he had attacked fishing villages in Nova Scotia, captured more than a dozen ships and generally disrupted the British war effort at every opportunity.

As the conflict progressed, Jones's unique talents were deployed to wreak havoc off Britain's shores under French naval command. By the time he had reached the Forth in September that year, Jones had already spent several months at sea fulfilling his mandate for provocation and chaos, seizing a Royal Navy fourteen-gun sloop, the HMS *Drake*, off the coast of Ireland, and undertaking a daring amphibious assault on the town of Whitehaven in Cumbria. Dunbar's impromptu show of force, however, must have worked. Jones was brazen but not reckless, and obviously thought better about engaging with what appeared to be a well-defended town. Instead he sailed west, heading up the Firth of Forth with a larger prize in mind.

Several days earlier, Jones had captured a collier which had sailed out from Leith. He had learnt from its captain that not only was Edinburgh's port poorly protected, but that several fine ships – a customs boat and three cutters – lay at anchor there. For someone of Jones's predatory mindset, the opportunity was too tempting to ignore. He formulated a plan to launch an attack at dawn: the harboured ships would be overpowered and a vanguard of troops would then be sent ashore to find the town's provost and dictate the terms of a ransom. Jones was as meticulous as he was audacious in his strategy, penning a courteous and deftly menacing letter to be delivered by the commander of his marines.

'The British marine force that has been stationed here for the protection of your city and commerce, being now taken by the American arms under my command, I have the honour to send you this summons by my officer,' wrote Jones with such impeccable manners, it may not have been immediately clear to the town's authorities they were being threatened. 'I do not wish to distress the poor inhabitants,' continued Jones in cordial tones, 'my intention is only to demand your contribution towards the reimbursement which Britain owes to the much-injured citizens of the United States.'

But then came the ultimatum, the iron fist in the velvet glove. 'Leith and its port now lies at our mercy; and did not our humanity stay the just hand of retaliation, I should, without advertisement, lay it in ashes.' It was a shakedown, a barefaced hustle. Despite the references to political motivations, Jones and his men wanted money. Their plan was to extort £200,000, promising that 'no further de-embarkation of troops will be made . . . and that the property of the citizens will remain unmolested', providing, of course, that the town's provost complied with their demands.

The letter, though, was never delivered. By 16 September, Jones's flotilla had reached Inchkeith Island and lay anchored off its shores waiting to attack. The ships were clearly visible from either side of the Forth, prompting fear-stricken responses from the local populations. In Edinburgh, pipes and drums sounded warnings of the imminent danger. Women and children were evacuated from the city, while men armed themselves with pikes and claymores.

At the same time, the townsfolk of Kirkcaldy on Fife's coast opted to place their fate with more celestial powers, mounting a form of pre-emptive spiritual retaliation. A local minister, the Reverend Robert Shirra, led his congregation and a gathering crowd to the beach at Pathhead Sands. Once there, he promptly took a seat and began fervently praying that 'the piratical invader Paul Jones might be defeated'. If local legend is to be believed, then the cleric's petitions were promptly answered. On the morning of 17 September, when Jones was 'almost within cannon-shot of the town' and with 'everything in readiness for descent', a sudden storm blew in from the west. The gales were so severe and unexpected that the would-be invaders were temporarily prevented from reaching the port and were stalled long enough to reconsider their mission, their captain then concluding it to be 'impossible to pursue the enterprise with a good prospect of success'.

By the skin of its teeth, Leith had been spared. But whether the town's escape was simply down to meteorological good fortune or, as the people of Kirkcaldy may have believed, the result of divine intervention, the episode marked a curious historical footnote and a salient lesson in the strategic vulnerability of the Forth. Jones's marauding interests remain perhaps the closest Scotland has ever come to officially being invaded by a military force of what would become the United States. More pertinently, at the time it highlighted the ease with which one of Britain's most commercially and militarily important waterways could be breached.

Unsurprisingly, in the wake of the aborted raid, a raft of defensive structures was built along the Forth coast. In Dunbar a battery of sixteen guns was erected on Lamer island at the edge of the harbour. Leith hastily constructed a fort which was occupied with an infantry detachment from Edinburgh Castle and later installed a Martello tower situated in the docks. Further up river, North Queensferry and the small island of Inchgarvie were heavily fortified with battlements and troops. Astonishingly, though, during this period and through the rising threat of the Napoleonic War, armaments were never installed on Inchkeith Island. Without knowing the exact reason, it appears a startingly naïve strategic oversight. Inchkeith lies at natural choke point in the Forth. Five kilometres north of Edinburgh and three kilometres south of the Fife coast, the island commands a central position where the estuary's waters narrow inland to the west and widen to the North Sea in the east. The long outcrop creates and guards two sea channels, funnelling the routes of all ships wishing to pass either side of it.

The tactical value of the island was also historically well established. In 1547 it was captured and held by a garrison of English soldiers during the brutal conflicts of the 'Rough Wooing'. Its protected position in the Forth, according to a nineteenth-century account in *The History of Leith*, afforded the occupying force 'many advantages, and they committed considerable depredations on the shores of Mid-Lothian and Fife, securing themselves from pursuit by returning to the island upon any alarm, where they were out of all danger from sudden reprisals'.

The repeated damage inflicted by the island-based raiding party eventually forced the Scottish army and its French allies to mount a response. A battalion of over a thousand French troops sailed out from Leith, deliberately making the journey in a number of small boats to conceal their attack. As they drew closer, however, the difficulty of landing discreetly and safely soon became clear. 'Nature itself has fortified it,' reported one of the French soldiers later, 'the access is so difficult that it cannot be come at except by three fit places . . . On all sides nothing is seen but continued precipice.'

But land they did, and a bloody battle ensued, the French troops coming up 'against eight hundred English trained to war, and accustomed to slaughter'. The skirmish was brief, but hard-fought on both sides, with over 300 men killed in the fighting. Inchkeith was seized by the Franco-Scottish force, and for almost twenty years afterwards it was occupied by the French, who reinforced it with a defensive citadel built at the top of the island.

It became a sea-bound fortress, easy for an incumbent force to protect and almost impossible for an invading army to capture. Control of the island also meant control of the Forth – a distinctly unsettling notion for the Scottish parliament. Fearful of a foreign power maintaining such a key strategic foothold so close to Edinburgh, the government requested that the French leave. Not only that, Inchkeith was put beyond military use, its defences dismantled so they could never be repurposed by Scotland's enemies. For the next 300 years the island remained largely unfortified, often repurposed in usage, but generally forgotten about. All that would change though in the final decades of the nineteenth century. Ratcheting geopolitical tensions between European nation states and the impending threat of global war would once again turn the island into a place of unignorable importance.

Conventional wisdom among sea kayakers dictates that even in the calmest of conditions you never go out on the water alone. Two people is the minimum, but on longer ventures, three or more is best. I was making the journey with two others, Alice, a local sea kayak guide, and Katerina, one of her clients.

We kept a steady rhythm, pacing ourselves in the heat. Halfway out with five kilometres still to go, the paddling became almost hypnotic, and I experienced that strange dislocating feeling which comes with repetitive movement in an unchanging environment. Every so often, though, something would happen to break this meditative state. At first, a small squadron of puffins flying in formation low across the water – a glimpse of plump bodies on frantic wings and the bright colours of clownish beaks. And then, later, a seal's snort from close behind. I turned to see wet, doleful eyes watching me intently. Seals are curious by nature but prefer to avoid eye contact, so will often follow kayaks, swimming around them but surfacing at the stern.

The third time we stopped was out of caution, Alice signalling us to wait. We were close to the shipping route that runs south of Inchkeith, and a kilometre or so away a fishing vessel was heading towards us, gunning its engines. We stayed still, bobbing on the water, trying to judge its speed and direction. I wondered momentarily if the sunlight on the water would absorb our outlines, consuming us in brightness so that we'd be all but invisible to the skipper. The boat, though, must have seen us. The sound of its engines dropped a couple of octaves, and at the same time its bow came around in a change of course.

We drew closer to the island, heading for a break in the cliffs at its western end. There was a harbour wall there, crooked like a curled finger to enclose a shallow area of light blue water. Above the bay and across the island I could see dozens of buildings: flat-faced rectangles made from pale breeze-block walls that stood out as abruptly as stage props against the island's backdrop. Smells drifted out to us – the acrid stench of guano and the fish odour of basking seals. As we landed, there was another sensory assault – a grating white noise of gull shrieks. The birds were immediately unhappy with our presence and began swooping, taking it in turns to carry out an orchestrated aerial bombardment. Each one performed a low screeching dive at our heads then pulled back up, lining themselves up for a return run.

Stepping on to the island felt like entering a demilitarised zone. The beach was an amalgam of pebbles and industrial-grade rubbish. Broken glass, steel cabling, blue plastic tubing, a section of chain-link fence, a tyre, and engine parts I couldn't identify. It was as if we had reached somewhere of recent rather than ancient conflict, surrounded by the detritus of massed activity and frantic departure.

From the shore, we followed a vehicle track that banked steeply uphill, passing a series of small, red-bricked huts that dotted the slopes on either side. The buildings were flat-roofed and mostly windowless, showing a murky darkness beyond their empty door spaces. Passing by them, it was hard not to harbour an irrational feeling of being watched, as though we'd inadvertently wandered into the first few frames of a slasher film. I started imagining our group being malevolently watched from inside the buildings and had to tell myself to get a grip. My next discovery though, made things worse.

The path, I noticed, was strewn with small, white sticks. Curious as to what they were, I stopped and looked closer. Not sticks, but bones. Hundreds of them, littering every section of the path ahead of us. They were familiar looking too. Drumsticks and breastbones, identical remnants to those of a Sunday lunch. I was completely baffled by their presence and sheer number. They were scattered everywhere, as if the place had been witness to some large-scale chicken slaughter.

My paranoia then suddenly increased. I became convinced that I was being followed. I looked behind to see an empty path, but as I continued uphill, the feeling remained. I turned around again, and still saw nothing. This was ridiculous, I thought; I was becoming as jumpy as a hysterical teenager. I persuaded myself there would be a logical explanation, and I speculated if some peculiar confluence of landscape and subconscious factors were at play. The dense vegetation, the mysterious bones, and the breakers-yard desolation of the island had set my imagination whirring, and perhaps these were somehow conspiring to suggest another's presence.

For whatever reason, I had to glance back a third time. On this occasion, though, I saw my stalker. Walking straight towards me was an implausibly large, russet cockerel. For several seconds it strutted confidently in my direction before nonchalantly ducking back into the bushes. I laughed out loud, perplexed by the unexpected sight, but also relieved that my sanity was still intact. At the top of the hill I met with Alice and Kat. They'd had a similar close encounter and were now surrounded by three or four hens and another rooster. The island had been last inhabited in the 1980s – permanently vacated when the lighthouse keepers left. The feral birds, I guessed, were the descendants of the keepers' domesticated brood and had managed to live wild here ever since.

We decided to explore in different directions, and I stayed on the track that ran along the spine of the island heading north. To my left the land fell away sharply, forming the island's north-western shore. A small bay came into view, nestled between the crags. In its centre sat a row of cottages, almost swallowed by undergrowth with white-flowered canopies of elder bursting through holes in the roof. A similar tangle of shrubland crowded the path around me. A throng of thistles, nettles and rosebay willow herb. 'Disreputable plants', the nature writer Richard Mabey affectionally called these waste-ground specialists, adept at growing in the margins and transition zones of human existence.

The track faded away, dissolving into a mass of thuggish vegetation. Ahead of me, at the highest point on the far edge of the island, was a blockwork building, stacked in haphazard sections. A lower level was accessed by a concrete stairway that ran to an underground bunker. Above that was a middle tier with a single doorway, and on the top was a sentry box, jutting out like a balcony with viewing slots on each of its four sides.

I explored below first, descending steps that had become compacted with earth and nettles. At the bottom were two rooms on either side. I went inside the larger, edging past a wooden door that had fallen off its hinges. Inside was a large tunnelled space that ran for thirty feet or more. The air within felt thick and smelt strongly of fungus. Light spilled in from a floor-to-ceiling window. The panes had long since shattered, leaving just a grid of empty mullions which threw a distorted geometric shadow on the floor. On the wall nearest to me was a row of wooden boxes, fixed as a shelving unit. Inside each space, a mulch of dried plant matter had accumulated over the years, left from the countless times that birds had nested in them.

Along the tunnel was a raised platform supporting

a set of metal tracks: a means of transporting heavy or delicate items underground. I knew that among the many uses of the buildings on Inchkeith, several had been specially designated for storing munitions. Magazines were distanced from other structures to limit the risk of damage on explosion. Gunpowder stores were particularly dangerous places. The likelihood of accidental detonation was so high that naked flames were prohibited anywhere near the rooms, and workers had to remove any metal objects before entering to avoid the chance of producing a stray spark.

At the far end, I could see daylight shafting in, revealing another set of steps leading to an exit. I wanted to see more, but in the darkness and alone I thought better about venturing further in. Back above ground, the sun was blinding. I cupped a hand above my eyes and scrambled up the broken brickwork of a half-collapsed external staircase. On the upper level there was an empty doorway leading to an enclosed viewing platform. The roof sloped down towards a concrete balustrade, leaving just a narrow slit and a protected field of vision that felt like being inside a bird hide.

The vista had been faultlessly calculated. Sea and sky had been separated in equal portions, the horizon exactly centred. The framing was almost cinematic: the coast of Fife banking out to the left, the wide blue sweep of sea, sky fading to white in the distance. Gulls swamped the foreground, carving the air in fast, cross-diagonal dives. Further out there was slower movement. Tankers and cargo ships idled across the main shipping lane north of the island. Closer to shore I could also make out a collection of smaller boats, fat-hulled things that trundled through the water like bath toys. Each plied their own small quadrants of sea, moving between marker buoys, checking their lines. Watched in widescreen through the observation tower, the scene seemed oddly staged, unreal and commemorative, as if re-enacting something once commonplace but now long since gone.

Less than two centuries ago, the Forth was filled with the fluttering umber sails of hundreds of wooden clinker boats. Villages right along the coast put crews out to sea: from Buckhaven, Largo, St Monans, Pittenweem, Ansthruther and Crail to the north, and Newhaven, Leith, Musselburgh, Fisherrow, Cockenzie, North Berwick and Dunbar to the south. They would row from the small harbours – tall-masted flotillas of drifters, yawls and fifies – all jostling for space until they reached open water. Once there, they would hoist four-cornered lugsails, catching the wind and sailing fast to the fishing grounds of the Traith, the Haikes and Peffer Sands, as well as the rich waters near the Isle of May and off Inchkeith.

The fishing fleets were in rude health. The Forth was a place of immense marine abundance, seemingly able to sustain any number of small-scale fishing enterprises. A bounty of inshore fishing was to be had within sight of the coast: haddock, cod, turbot, skate, whiting, mackerel could all be caught within hours of leaving port and landed the same day. 'There are great quantities of white fish taken and cur'd upon this coast, even within, as well as at the mouth of the Firth,' reported Daniel Defoe in 1727, noting even then the scale and reach of the industry, 'I took notice the fish was very well cur'd, merchantable, and fit for exportation;

and there was a large ship at that time come from London, on purpose to take in a loading of that fish for Bilboa in Spain.'

As well as the fish, there were oysters. Vast beds stretched out from Hound Point in West Lothian to Gosford Bay in the east and north to the towns of Aberdour and Burntisland in Fife – an area of fifty square miles, yielding as many as thirty million oysters a year. A wider ecosystem of larger fish and mammals was also supported. Several species of whale, including orca, were regularly seen in the Forth and were often taken by fishermen, highly prized for their oil. Blue whales too were occasionally recorded entering the estuary, with the species first scientifically described by Sir Robert Sibbald from a specimen washed ashore in the Firth of Forth in 1692. Seals, dolphins and porpoises were also profuse, and were hunted opportunistically rather than systematically, killed largely because of their perceived impact on fish stocks. But despite all the ecological diversity on offer, the lives and fortunes of the Forth's fishermen revolved around the two annual migrations of one particular fish.

Every year, with almost unfailing predictability between late July and early September, herring would arrive in their millions to spawn in their traditional breeding grounds along the Forth. Coinciding with the pagan festival of Lammas, the influx of fish and the mass excursions to capture them became known as the Lammas Drave, a bonanza so huge and so reliable that local boats were joined by fishermen from as far away as Wick in the north to Lowestoft in Suffolk, as well as Dutch fleets from across the North Sea. The Lammas Drave and the other annual herring migration, the Winter Herrin', brought much-needed prosperity to the region, an economic multiplier effect cascading wealth far beyond the pockets of fishermen.

The successes of each season's catch were centrifugal, rippling out to benefit a vast number of livelihoods and professions. Shipwrights, sail-makers, net-tanners, joiners and net-makers were all highly demanded trades, essential to keep the fleets fishing. Behind them was a whole host of ancillary occupations – the makers of baskets and fish boxes, curers, the operators of salt pans, even coopers who made the herring barrels – each directly profiting from the cyclical shoals.

Entire families were involved in the process, with jobs allocated strictly according to gender. While fishing was the exclusive preserve of men, it was women who would provide the labour vital for the industry's commercial success. Gutting, packing, mending nets, baiting lines and hawking the produce were all roles carried out by the female population. 'So laborious, essential and particular was the work of women in fishing communities', observed the historian Christopher Smout, 'that it was rare for the men to marry outside them.' The fishwives were indeed an indomitable tribe. The men may have braved the dangers of sea, but it was the women who had the most arduous task of all; 'on four days a week they carried the fish in creels for the five miles into Edinburgh, sometimes bearing the fish in relays with three women to one basket, when they could reach Edinburgh in three-quarters of an hour.' Most impressive of all were the women of Fisherrow who would accompany their husbands to Dunbar for the Lammas Drave. From there they would carry baskets of 150–200 fresh herring all the way to Edinburgh, a journey of over twenty miles.

Of course, the binary outcomes of such an all-or-nothing economic monoculture brought huge risk. The intricately connected framework of jobs and prosperity that were supported by the Lammas Drave and the Winter Herrin' all hinged on the fish returning in plenty each season. In the years when this didn't happen, the effect was devastating. At times, one poor migration would follow another, and the lean years would merge into periods of deep austerity – the 1720s, the 1770s and 1820s were remembered not only for the lack of fish but also for the poverty that resulted.

History, though, provided a degree of reassurance. Each time the catches had previously slumped, the shoals had eventually come back as plentiful as before. So, when herring stocks suddenly declined following the largest Lammas Drave on record in 1860, it was assumed, as always, that they would soon return. This time, however, it was different. Years turned into decades, and the once dependable arrival of the summer shoals never materialised; the Lammas Drave was gone forever.

A combination of factors was to blame, with overfishing the prime culprit. By the mid nineteenth century, fishing in the Forth had reached an unprecedented scale. The boom years had enabled much bigger boats to be constructed, some up to forty feet long and twice the size of traditional vessels. Larger nets with smaller meshes also began to be used, expanding catch sizes, which now included more immature fish. In addition, fishermen increasingly began to ignore long-standing religious observance and go out on the Sabbath, thus depriving the herring of one day in seven to spawn. For a fish population that was about to reproduce, it's likely these kinds of pressures were ecologically disastrous, eventually reducing their numbers to a point where it was impossible to reverse the decline.

For some time afterwards, the Winter Herrin' still appeared, but the lessons of the Lammas Drave had not been learnt. Fishing became even more industrialised, with steam-powered drifters replacing sailing boats and the new and destructive technique of seine-net trawling accelerating the race to extinction. By the middle of the twentieth century, the Winter Herrin' also stopped. The annual migrations of herring to the Forth, an ebb and flow whose natural cadence had existed long before the arrival of humans, was suddenly extinguished.

The removal of such a keystone species also had wider implications on the Forth's marine wildlife. Haddock, codling, turbot and flat fish which predated the herring and their spawn would all be impacted. In turn, the depleting effects of species loss radiated further up the food chain, depriving birds and mammals of their food sources, and leaving an ecosystem profoundly and permanently altered by human activity.

The ground ahead of me sloped away rubble-strewn and blanketed with thick undergrowth – a terrain of unseen hazards. I decided to double back and headed instead towards the lighthouse and the island's highest point. A dry wind had lifted from the west, filling the air with a dusty light. I had to squint to read the landscape, picking out buildings wherever I looked. Each one was as severe and angular as the next, harsh geometry caught in an arid, pressed-down brightness.

Near where I had first walked up, I cut west to reach a tall, honey-coloured tower. It resembled a belfry – thin and imposing and designed solely for height. It was built on top of an older fortification, a large retaining wall of dark stone masonry that banked against a swell in the hillside, flanking it neatly on either side. An entranceway led into the protective depths of the fort, but I chose to stay outside and circled up above the ramparts. Up close, the tower had an end-of-civilisation look about it. Abandoned, but not yet decayed, as if lingering in the aftermath of some unspecified cataclysm. The doors and windows had disappeared, but wiring conduits and lights fittings remained fixed on the brickwork. A metal staircase also curled round the outside, leading all the way to a fourth-storey room, a viewing gantry with horizontal window spaces that cast narrow-eyed glances seaward.

Further round, I came across a large circular construction; a ring of concrete at least fifteen feet in diameter comprising a thick outer wall and a smaller circular pit sunk below the upper platforms. The descending levels and the recessed inner space gave the impression of the setting for some kind of small-scale open-air performance. It was in fact a gun emplacement, emptied of its armament and now oddly sculptural in appearance.

Before journeying to the island, I had read about these fortifications in an obscure nineteenth-century periodical, called *Edinburgh Old and New*. 'The guns are placed on a granite platform, in the centre of a circle, formed by a bomb-proof parapet', wrote the author James Grant, then explaining usefully the cannons were to be 'fired *en barbette* over the slope and not through embrasures, as they are worked on the Moncrieff swivel principle, which permits them to be turned so as to sweep any point within three fourths of a circle'.

As well as expressing incredulity that Inchkeith had lain undefended for so long, the periodical also referenced some of the island's previous non-military functions. According to an order from the privy council of Edinburgh in 1497, victims of an infection referred to as 'grand-gore' were to be removed from the streets of the city, assembled at the sands in Leith and then transported to the island, where they were to remain 'till God provide for their health'.

The disease was most likely to be syphilis, which had first been recorded only a few years earlier in Naples, but which had then spread rapidly throughout the rest of Europe. With no known cure, the sufferers are likely to have endured a slow and painful death on the island. 'There, no doubt,' Grant surmised, 'many of these unfortunate creatures found their last home, or in the waves around it.' This was geography employed as a means of social cleansing. A landscape close enough for the contained relocation to take place, but remote enough to deliver a comfortable moral amnesia for Edinburgh's civic leaders.

Inchkeith's proximity to, but relative inaccessibility from, the mainland meant that in the centuries which followed it found repeated use for the same purpose – a place of quarantine-cum-imprisonment. Edinburgh's plague victims were purged there in the sixteenth and seventeenth centuries, and disease-stricken ships bound for the capital were commanded to berth on its shores. On one occasion in 1799, records show it became first a lazaretto, then soon after the burial place for the afflicted crew of a Russian vessel.

Banishment to the island had also once taken a stranger, though no less callous, form. According to the sixteenth-century Scottish historian Robert Lindsay of Pitscottie, King James IV had reputedly exiled two infants to Inchkeith in the care of a woman who was deaf and non-verbal. Once there, deprived of linguistic input, the monarch believed that whatever language the children acquired would be mankind's innate, God-given tongue.

I wondered if knowing about these grim collected histories had somehow preconfigured my perception of the place, had in some way determined the feelings of apprehension and wilful detachment that I was increasingly registering whilst being there. This was no island of happy marooning. And the more I pondered the occurrences of its past, the more I sensed a black emotional undercurrent intruding on the present. 'I had not thought to encounter these older darknesses,' wrote the author Robert Macfarlane, explaining how interpretations of landscape and history can often be so close as to be inseparable. 'I had passed through lands that were saturated with invisible people, with lives lived and lost, deaths happy and unhappy, and the spectral business of these wild places had become less and less ignorable.'

But perhaps the effect also worked in reverse. Perhaps, as in the words of the writer Rob Cowen, certain places were capable of some kind of 'emotional transference', that 'landscapes somehow become repositories of personal and collective memory', that 'traces are imprinted or stored in an imperceptible or intangible way and the land itself retains the culture of a place'.

'It seems that even islands like to keep each other company', wrote D.H. Lawrence in an aphorism which seemed entirely plausible the more I found out about the islands of the Forth. They were indeed familial – linked by shared geologies under the water and by corresponding histories above the surface. Most were first revered as places of spiritual sanctity, hermitages of saintly patronage which together formed a constellation of religious significance along the Forth.

Inchkeith was supposedly inhabited by St Adomnán, while the splendidly named St Baldred of Tyninghame was said to have settled on the steep volcanic plug of the Bass Rock. The cells and monasteries were well known, and as a result were often targeted by raiding parties. In 875 AD the Isle of May – the most easterly island in the Forth – was the subject of a particularly brutal attack by Danish Vikings. The marauding Norsemen landed on the island and are said to have massacred its entire ecclesiastical community, including its figurehead, St Adrian. Similarly, the island of Inchcolm, home to a strikingly beautiful Augustinian abbey, was plundered by English forces during the Wars of Scottish Independence. In a story which echoed the one I had heard about St Finnan's Isle, Inchcolm's treasures had also seemingly been afforded godly protection. As the English sailors made good their escape with the stolen artefacts, they apparently became so besieged by bad weather they almost wrecked their ship on the skerries of nearby Inchkeith. Believing this to be a direct result of their misdeeds, they turned around and returned their ill-gotten gains, after which the sea was once again becalmed.

There were also recurring themes of usage among the islands. Like Inchkeith, both Inchgarvie and Inchcolm were employed as quarantine sites. The former was also used as a place of incarceration, as was the Bass Rock, which had once been the setting for one of Scotland's most notorious prisons.

It was here, following the restoration to the throne of Charles II and his subsequent persecution of the Presbyterian faith, that hundreds of Covenanters were gaoled in the Rock's castle. By all accounts, it was a truly wretched place. Convicts were held in cold, dank cells, permanently chilled by exposure to wind and sea-spray, with rations of food and water so limited they captured seabirds for nourishment and drank rainwater from puddles. Control of the island, though, changed hands seventeen years later when the Catholic James VII was deposed. The Covenanters were released and the fort was used instead to hold four young Jacobite sympathisers in what turned out to be a famously brief imprisonment but otherwise lengthy stay on the island.

When a supply ship was being unloaded by the castle's guards, the four prisoners saw their chance to outwit their captors. Having escaped from their cells, they closed the prison gates, turned the guns on the hapless gaolers and ordered them to leave. Sustained by regular supplies smuggled out to them, the Jacobites then held the island for almost four years, until an exasperated British government blockaded the rock. When the Jacobites eventually signalled a truce, they invited the commanding government officer to dine with them and agree the terms of capitulation. To his astonishment, he was treated to a lavish feast complemented by the finest bottles of French brandy. Unbeknown to the government officer, these were the final supplies left on the island but were deliberately consumed with such casual abandon he was convinced there were provisions there to last for years. As a result, the Jacobites negotiated terms of surrender for more favourable than if the truth of their situation had been known and were granted safe passage to France or the freedom to remain in Scotland.

As well as being boldfaced, the besieged men had also been lucky. For despite appearing to be uninhabitable, the island they had seized was home to an almost inexhaustible food supply. Then, as now, the Bass Rock teemed with gannets, a colony so large it transforms the appearance of the island each year. From early spring, when the birds first arrive, until September, when they leave, the island's colour is altered, shifting from the subdued greys of its winter rockfaces to become bright white, a compound change in hue caused by the massed nests of breeding gannets. Seen from the mainland, the effect is striking; within weeks, the volcanic dome is rendered with an Arctic guise, turned from island to iceberg.

For centuries, birds had been hunted from the rock, for their meat, feathers and fat. Young gannets were particularly sought after and were subject to so much human predation that in 1592 they were safeguarded by one of the earliest pieces of British conservation legislation, preserving the economic value of the birds for the sole benefit of the Bass's owner. Exploitation of the colony though – most notably through recreational killing – continued until the early twentieth century, when full legal protection was granted for the birds throughout the breeding season.

The impact was immediate and astounding. When not faced with indiscriminate slaughter, the population expanded at a rapid rate, rising from around 3,000 pairs in 1904 to almost 40,000 by the end of the twentieth century. Over 100,000 gannets now nest on the island, which roughly equates to 12 per cent of the bird's global population, making it the largest gannetry anywhere on the planet.

The phenomenal success of the Bass gannets was mirrored

by the regeneration of bird species on the other islands of the Forth. The Isle of May, the largest island on Britain's east coast and now one of the country's most important bird sanctuaries, has witnessed startling revivals. Populations of fulmars, terns, cormorants, shags, kittiwakes, guillemots and puffins have exploded in the last hundred years. And although grave threats still remain, most pertinently from the decline in sand eels (the staple diet of many of the species), bird colonies on the Forth's smaller islands such as Fidra, Craigleith and the Lamb have also thrived. It has meant that amid all the islands' dark and often militarised histories, a separate narrative has evolved. They have become places of non-human stronghold, redoubts of a natural kind.

A small townscape converged at the pimpled summit of the island. A line of terraced houses stacked upwards on the elevated ground, bordered by concrete slopes and a stout stone wall. Beyond that were more square-faced military buildings crammed one above the other, and higher still was the lighthouse, its stubby central tower painted a curdled-cream yellow.

The rising architecture, lofty and self-contained, reminded me of other pinnacle buildings – the *villages perchés* of Provence, or more dramatically, the refuges on the Aiguille du Midi in the Alps. In an alternative reality, the place could have been quaint, beautiful even: steps winding between the buildings and cliff-top views above the bay. As it was, though, I struggled to see past the decades of decay and the eerie emptiness that seemed to accompany it.

At the top, there was a courtyard. A dusty open area, criss-crossed with a grid pattern of weeds that had grown rampantly between the paving cracks. The gulls became even more vexed as I entered the space. I was swooped on and hit twice, and I broke into a half run to reach the lighthouse. Against the other buildings, it appeared an awkward contradiction of preservation and colour. The windows were all intact, covered with black security grills, and a row of solar panels had been festooned along the railings of the light's balcony. A bright metallic lightning rod led from the top of the dome and next to it aerials were fixed, pointing skyward. Above the entrance, the building's commemorative plaque was still legible. 'For the direction of mariners, and for the benefit of commerce. This lighthouse was erected by order of the commissioners of the Lighthouses', it stated in a fittingly grandiose tone. Underneath was the date it was first illuminated and the name of its engineer, Thomas Smith, the patriarch of the Lighthouse Stevensons.

To my right was an archway leading to a small close overgrown with elder. There was movement beyond, the bushes trembling as chickens skulked about in the undergrowth. I noticed a stone tablet placed above the opening, a crest displaying what looked like two unicorns rearing up on either side of a shield with a lion rampant, and over them the motif of a crown.

I had heard about this carving, but was unsure if I would be able to find it. The panel had once been part of the original fort and referenced a visit of Mary, Queen of Scots to the island in 1564, the date still visible below the royal coat of arms. At some point in the razing of the first citadel, the stone had been salvaged, its significance acknowledged, and it had found its way back into the fabric of the later building.

I wandered among the derelict buildings, peering rather

than venturing inside, not trusting that the green floorboards would take my weight. Most of the rooms had the frozen-in-time look of museum exhibits, neglected but still with their original features. Wooden cabinets were fitted to walls, coat hooks on doors. There were light switches, shelving units, and the elaborate mosaic of floor tiles. Like the wrecks of abandoned whaling stations or the interiors I had seen of deserted gold-mining towns, they appeared as strange tableaux, snapshots of when their usefulness had finally ceased, captured in the very moment of their irrelevance.

Away from the cool shadows of the summit buildings, I could see Alice and Kat moving on a lower part of the island's ridge. They walked between more ruins, their outlines dark and fuzzy in the heat. It would take months, I thought, to reconnoitre every one of Inchkeith's buildings, longer still to understand the sheer complexity of how the island had been used. Almost 600 men – the population of a small village – had been stationed here during its most intensive period of active service in the early years of the Second World War. And with such an occupation came the requirements of significant infrastructure.

Among the buildings I could see were the remains of storehouses, workshops, offices, drill halls, billets, latrines and even a field hospital. Fully operational, the place would have been exceptionally busy, bristling with activity and primed with the nervous anticipation of conflict. Gun emplacements, searchlights and observation posts were positioned along each side of the island. Trenches and pillboxes providing small-arms cover looked out from every defensive exposure. Where cliffs could be breached, barbed-wire entanglements were strung above the tideline and floating booms and mines spanned the surrounding water.

The defences were undoubtedly necessary. With the navy's most important shipyard situated at Rosyth and with the harbourage of many of the British Home Fleet's ships in the estuary, the prospect of military engagement was likely at any time. As well as the threat from aerial bombardments, U-boats regularly infiltrated the Forth and were a constant and concealed menace, responsible for sinking a significant number of vessels.

Of equal danger were mines. Ships would regularly stray into unswept areas or simply run out of luck, coming into contact with floating ordnance and triggering devastating explosions. Some of the worst incidents occurred off Inchkeith, including the sinking of an Admiralty tug in 1940 with the loss of twenty-seven lives, and the mining of a cargo ship the same year in which ten men died.

Another boat was spared a similar fate when gunners in the southern battery of Inchkeith spotted the trawler sailing dangerously close to a minefield. Unable to raise the alarm by radio, the artillery men fired a practice round across the ship's bow. The warning must have worked, but its method of delivery could have proved fatal. Having safely missed the vessel, the shell bounced on the water and continued on towards Leith, where, several kilometres later, it smashed through the wall of a tenement building before coming to rest in a garden shed. Incredibly, no one was injured, and according to some retellings, the projectile was politely mailed back to the garrison on the island.

It was almost time to go, and I felt ready to leave. I had only seen a fraction of the island, but I had been dislodged by it, knocked off centre by the baleful echoes that seemed to reverberate around the place. I was being irrational, I knew: unable to disentangle an emotional response from the knowledge of its history and the brutalist aesthetics of its present. But in the same way as other landscapes might provide comfort or a benign wonderment, Inchkeith seemed to me to exert the very opposite, an almost palpable sense of disturbance.

Alice and Kat were waiting by the kayaks when I arrived. It was late afternoon and the sun was still strong, casting an angled light and edging the island's buildings with dark outlines as I looked back from the water. The wind was up as we paddled home, a following wind gusting behind us and lifting the sea in small waves. I thought a lot about the island on that journey back. I had never visited a place where the contrast of wildness and human utility had felt so forsaken. I wondered how we accommodate and account for such landscapes, if indeed they could ever be properly understood. Perhaps they never can. Like Inchkeith and the other islands of the Forth, there are wild places with wild histories that defy categorisation, that have flickered in and out of relevance over the centuries, becoming layered with different strata of physical and narrative remains. I wondered as well about a time when Inchkeith might become important again, what factors would reignite its significance, and what larger meaning this would have.

It took a couple of hours to reach the long arc of sand at Portobello's seafront, the shore from which we had left. By now it was early evening, and the beach was busy as we approached – people freed from their working days, families and friends grouped together. From a distance, there was the smell of barbeques, and I could hear the excited yells of children playing in the water.

The Bone Caves

Inchnadamph, Assynt

Ever since the night I had spent in Glen Loin, canopied from the winter stars and held safe in its warren of hill-side boulders, I had wanted to search for other cloistered landscapes. Places whose relevance was determined not by what had happened on them but rather what had occurred within them. Caves, hollows, caverns – all exert a forceful gravitational pull on me. Something atavistic, deeply primal, but also something intensely curious. I'm a sucker for their mysteriousness: the enigmatic potency of truth and myth which all below-surface spaces seem to harbour.

Partly it's their form. An inverse world, inwardly expansive but outwardly indiscernible, possessing neither shape nor outline of their own. They are topographic illusions, hidden within contours, profiled seamlessly to the upper realms – being 'of' the landscape rather than constructed from it. Few places are therefore more ancient in their human association or more enduring in the continuity of their use.

As the earliest of shelters, they would have been lifelines. Ready-made refuges as durable and discreet as the ground they are embedded within. A bridging point between mere survival and societal development, places to buy time, live a little longer, a base from which to build. And perhaps this value has a heritable link, a genetically transferred appreciation that humans have always referenced and returned to again and again in times of hardship and persecution, hostility and upheaval.

By their nature, they have often also been secret places. Coveted, unshared and largely unspoken of. I knew of a handful of hidey holes and ground-lairs across wild areas of Scotland. Caves, mainly, their names attributed to a one-time resident, their location reaching me only by word of mouth. But there are many more. Some have even gathered a cachet of celebrity. The multiple, supposed coverts of Rob Roy MacGregor or Bonnie Prince Charlie can be found (sometimes signposted) in remote areas right across Scotland.

This profusion though is surprising and uncertain. Scotland is not endowed with the geology for caves, lacking an abundance of soluble, karstic rock areas. Yet these clandestine landscape features exist in numbers greater than can be imagined or accurately recorded, though some have tried. I have often come across the quietly shared knowledge of cavers and cave-lovers, and the scale of detail is astounding. Esoteric lists in out-of-print books and website directories note entrances and exits, grid references and rock types. Almost always, however, there are caveats. Acknowledgements to the limitations of the information, question marks over what has still to be discovered.

I have noticed too that a vague taxonomy exists for these places which seems to correspond with their level of concealment or perceptibility. Sea caves, the large wave-carved openings on coastal cliffs, are normally the most accessible and among the best known. Boulder caves, formed from voids in rock-fall, also feature prominently in Scotland and are often reachable but hard to locate. The most hidden though and the most dangerous are potholes, the vertical passageways to subterranean rivers and the deep underground.

Finding a cave then, whether by accident or by design, feels like one of the wildest connections to be made with a landscape. A touchpoint to somewhere of unfathomable antiquity, a veritable sinkhole into the past. So, on a bright, wind-filled day, we went exploring, my ten-year-old son and I – searching for caves.

We had travelled to Assynt, in the far north-west of Scotland, the last bulwark of mainland that edges out to the eastern waters of the Minch. Inselbergs – island mountains – rise here. Suilven, Quinag, Canisp, Stac Pollaidh, Cul Mòr and its sister Cul Beag mass like sandstone galleons, their shapes long and thin, steep prows cutting up from peat and moor. It's a landscape of emergence, of geological thrust. Rock pushing on rock, layering, stratifying. It is also, though, a place of reduction, fretted and weathered. Water steeps and reflects here. Not only by light but by action, eroding the mountains above the surface and carving the ground below so that it becomes a mirror landscape, vertically inclined across a horizontal plane, downwards as well as up.

Limestone is the reason. Pavements of the perforated stone – a rarity in Scotland – stretch in lateral sections between the harder more ancient Lewisian gneiss. Water works corrosively on limestone, dissolving its weaknesses and boring at fault lines. Tunnels are opened up, creating networks stretching deep into the Earth, making Assynt famous among speleologists. The small area of land running diagonally south-west to north-east from Knockan Crag, through the village of Elphin and on to Gleann Dubh, contains Scotland's largest and most extensive caves. Several major systems exist here – the longest stretching for over two kilometres – but new sections are still being found, its underground complexity still being unravelled.

At the tiny hamlet of Inchnadamph we turned off the road and into a large, empty car park, dust billowing as we crunched to a stop. It was mid afternoon, searingly bright and ferociously windy. Gusts tugged at our jackets as we loaded our rucksacks. Tiny flecks of dried grass filled the air around us, catching in our clothes and stinging our eyes. Light bounced off every slick surface: the paintwork of the car, the tinfoil water of Loch Assynt and the closed windows of the nearby Inchnadamph Hotel. We had arrived out of season, although it was hard to imagine the handful of buildings and its hotel as ever having a season in which it could be out of. But for over 150 years tourists and visitors have been coming to this place, lured by its geological uniqueness.

It was here in the second half of the nineteenth century that the battle lines were drawn on an acrimonious scientific debate. Assynt and Inchnadamph were to become the geographical focal point of the 'Highland Controversy', a bitterly partisan disagreement regarding the age and formation of rock types in the north of Scotland. It was an argument which would fester for several decades, but by its conclusion would upend pre-existing concepts of geology and establish new principles of analytical fieldwork.

In the mid-nineteenth century, the prevailing concept put forward to explain the structure of Britain's geology had centred around the work of eminent Scottish scientist, Sir Roderick Impey Murchison. His Silurian theory, developed in the mountains of Wales in the 1830s, described mountain-building by accumulation: metamorphic rock forged in the Earth's molten abyss between 417 and 443 million years ago was then thrust upwards as the planet's crust compacted. These actions, Murchison proposed, could be witnessed quite clearly in a linear ordering of rock types, stratified chronologically in terms of age – younger formations layered on top of older ones.

By the early 1850s, in an attempt to extend his Silurian interpretation to characterise a wider area of the United Kingdom, Murchison made a series of field trips to the Highlands, first assisted by James Nicol and later by Archibald Geikie. Murchison believed the research he had carried out in Scotland had proved conclusive. The Scottish landscape, he determined, fitted conveniently within existing ideas on the geological timeframe and formation that he himself had already observed and named elsewhere.

Others, though, were less sure. Nicol expressed reservations which quickly turned to doubt and eventually to outright dispute. His views were shared by a number of prominent but amateur geologists, most notably Charles Lapworth, who were at odds with the evidence presented and the apparent lack of scientific rigour behind the study. The problem was that there were rock formations in parts of the Highlands which just simply did not fit within the Silurian paradigm. At Knockan Crag in particular, Moine schists could be seen overlying rock that was hundreds of millions of years younger, and layered much further below. How, wondered many of the more enquiring geologists, could these anomalies exist and Murchison's theory still hold water? Murchison, of course, had an explanation. The ancient strata of gneisses and schists observed in Assynt sitting above rock types formed more recently were, in fact, not as old as they seemed. They were, according to him, simply newer metamorphic formations.

Two forces were at work during this nascent period of geological science that explain the intellectual hegemony of the Silurian view. The first was the powerful momentum of Murchison's own self-willed belief. His conviction in the strength of his idea and the deluding myopia of confirmation bias that must have seeped into his analysis.

This was also a time of academic empire building, of reputational advancement and political positioning, where a theory could hold sway through sheer force of persuasion and the strength of personality. And Murchison, indeed, had an almost incontestable establishment persona. A military veteran and cohort of geologists Charles Lyell and Adam Sedgwick, he had been elected a foreign honorary member of the American Academy of Arts and Sciences in 1840 and six years later was knighted. By 1855 he was appointed to the most pre-eminent position in geology, serving as the director-general of the Geological Survey (which later became the British Geological Survey).

To mount an opposing view, therefore, required not just robust evidence but also the social and intellectual gravitas to influence the geological community at large, its career frameworks and its peer-review structures. As professor of natural history at the University of Aberdeen, James Nicol was too marginal a figure geographically and politically to dislodge consensus opinion. Even after Murchison's death in 1871, prominent academics faithfully carried the torch of his theory, most notably his protégé Geikie, for whom Murchison had helped to the position of inaugural Professor of Geology and Mineralogy at the University of Edinburgh.

But a groundswell of challenge had been building. Slowly but surely, further research continued to question and undermine the application of the Silurian system to the Scottish Highlands. The painstaking fieldwork of the brilliant and largely self-taught Lapworth dealt the most decisive blow. While working as a schoolmaster, Lapworth had not only unravelled the structure of the Southern Uplands, his work had also led him to the identification of an entirely new geological period – the Ordovician. When he turned his detailed attentions to the north-west of Scotland, Lapworth was able to conclude that the existence of older rocks overlying those much younger was part of some enormous upheaval, a huge mountain-forming event so powerful it had churned the Earth's bedrock from great depths, angling and sliding it to the surface in a gargantuan act of orogeny. So powerful was this image, Lapworth even suffered from nightmares about being caught up in the folding crush of what he described as the 'great Earth engine'.

By the early 1880s Geikie's continued adherence to Murchison's 'official' theory began to seem increasingly precarious. Against the backdrop of compelling counter-evidence, Geikie became duty-bound in his new role as head of the Geological Survey to try and resolve the Highland Controversy once and for all and appointed two salaried geologists, Benjamin Peach and John Horne, to undertake the research. For over two decades Peach and Horne ranged over vast distances of the north-west Highlands, often from their base at the Inchnadamph Hotel, intricately mapping the complicated geological structure of the area. They were, as many colleagues observed at the time, the perfect foil for each other. Peach, powerfully built and leonine in appearance with an abundance of white hair, was instinctive and intuitive in his approach, prone to fits of frenzied activity as well as periods of lethargy. Horne, by contrast, had the bearing of an educated professional, neatly attired with an administrative proficiency and analytical diligence to match.

The work that the men carried out culminated in what is now regarded as a classic of regional geoscience. The publication, rather prosaically named *The Geological Structure of the NW Highlands of Scotland*, was to be far more impactful than its title suggested, for as well as closing the long-standing disagreements of the Highland Controversy, it helped completely reimagine the way the world's largest mountains were created. 'Rocks had been pushed over each other, slice after slice,' wrote Geikie in his magnanimous introduction and summary of the book, 'huge sheets of the very oldest masses having been torn up and driven westwards for miles.' By describing, for the first time, such colossal geological actions, Peach and Horne had visualised a process of mind-blowing scale: immense tectonic movements and deep-time collisions between continents that not only explained the enigmatic landscapes of the north-west Highlands but also shed light on the creation mystery of mountain ranges across the world. Assynt in turn became a Mecca for geologists, 'a classic region', Geikie noted, 'for the study of some of the more stupendous kinds of movement by which the crust of the earth has been affected'.

From the roadside, we cut between the hamlet's white-walled buildings, edging past neat-fronted driveways and heading upwards into a low-slung valley, the sky beyond a bright cerulean blue. We were following the River Traligill, walking in Peach and Horne's footsteps, treading the same route they would have paced out countless numbers of times. I thought about their journeys in this exceptional landscape, driven first by enquiry, then by analysis, and then eventually, I suspected, by love of the place. After their research was completed, and with the numbers of visitors to the area growing, Peach and Horne maintained their devoted connection to the area, acting as guides and leading groups up from the hotel to marvel at the geological features they had first identified.

My son moved ahead of me, always two steps in front. The rucksack I had given him was slightly too big, loaded with kit and sliding on his shoulders like an oversized coat. It made him look clumsy and committed all at the same time. On our right, the river appeared in glimpses: gasps of white water, or pools of blue stilling over grey rock. Its course, I had noticed from the map, was strangely intermittent, appearing and disappearing in some kind of cartographic conjuring trick. One minute it was there, the next its cobalt line had abruptly halted. The same was true elsewhere. Further into the interior where we were heading, I spotted a water system marked which was entirely independent of any others. A three-pronged confluence that just began and ended in the middle of the high moor.

Progress was slow. The gusts we had experienced lower down had now formed a constant stream of wind, funnelling between the valley walls and channelling an unseen current against us. Every few steps we were stopped in our tracks, held firm by the force of an invisible surge. In the distance, clouds tore fast across the summit of Conival, snagging threads of grey vapour that lingered like contrails before melting away seconds later.

We passed a lone, single-storey cottage of white brick and black roof, perched on a swell of higher ground and set about with hundreds of daffodils, their stalks bent double in the wind. Nearby, I spotted the cleaving run of a small dried-up burn, cutting through the hillside and meeting the path. I tracked its journey further down the slope until suddenly it shrank from view, burrowing into the landscape. At its vanishing point, we found a hole, two feet wide and bored into the incline. We cleared the debris that had meshed at the entrance and peered inside, seeing only blackness.

As the valley rose and small outcrops of milky limestone began surfacing among the heather, we saw the first cave. It opened up from within a grassy dome, a slanting archway that looked like the narrow glare of a dragon's eye. The roof was formed from lintels of curved rock, layered one above the other, appearing more architectural than natural in form. A small gap had been worn away by footfall at the entrance, and we stood at the edge looking in. The floor sloped steeply inwards – forty-five degrees of slick limestone devoid of handholds.

About ten feet down, at the base of the cave, was white water. A foaming cascade, luminous even within the cave's darkness, sliding fast and splintering noise up through the opening to give the place its name, Uamh an Tartair – the Cave of the Roaring. I wondered how many people had stepped beyond the threshold wanting to see more, intrigued by the strange velocity of the underground stream, curious as to where it went. I recognised a similar pull but also a lurch of fear that rocked me back on my heels.

We followed the path as it curved up and around leading to a separate entrance to the system, a wide crater with a vertical drop of twenty feet or more. At some point the ground here had collapsed, the roof of the chamber giving way to form the large opening. Small trees grew from the inside rim of hole, their trunks bending out at ninety degrees upwards to the light. I could still hear, but not see, the waterslide below. As in Glen Loin, I sensed a landscape only partially observed, the surface an unreliable membrane, ruptured and revelatory; ground that couldn't be trusted.

Further up the hillside, at the end of a short grassy trail we found a second cave. A trim semi-circle banked into the rock, covered from above by turf and moss. It looked homely, tucked beneath the level of the moor and nestled out of sight. I stepped inside and felt an instant calmness, the wind falling from my clothes, the air around me stilled. This was a place of natural shelter, of protection from even the harshest elements. I thought wistfully about what it would be like to visit in the depths of winter, imagining how inviting the place could be: the view looking out on to falling snow, a fire being kept, heat and light radiating within its curved walls.

Again, I could hear water, but this time only faintly. The sound was issuing out from a small hole in the base of the rock. I had read that a larger, secondary chamber existed on the other side, and it was through this cat-flap-sized gap that cavers would squeeze themselves to reach it.

When we resurfaced into the valley's airstream, the wind felt colossal. From there on we were pitched at a permanent slant, doubly inclined against the slope and the onrushing gale. We reached the top of a narrow gorge: limestone riven so steeply that we couldn't see the bottom, exquisite flutes and curves appearing in the rock. I walked with my hand permanently hovering over my son's shoulders, ready to haul him back if the wind pushed him too close to the chasm.

Eventually, though, it was all too much. As we topped out on the plateau of higher ground that we intended to cross, we were hit with the full strength of the strange, blue-sky tempest. My boy was knocked from his feet, sent skittering into the heather. I followed after, blown into the same small hollow where he had landed.

There was no way we would continue across the open moor that day, nor over the isolated peak of Beinn nan Cnaimhseag as we had planned. Instead, we lay there on our backs, side by side, clouds scudding fast above us.

Caves in their imaginative form often exist as they do in their physical form: below the surface and only partially perceptible, tucked away in the landscape of our cultural consciousness. When they appear, it is normally in our peripheral vision, indefinite and unconfirmed. In literature, their inclusion is frequently subtle in reference, albeit narratively significant: a wild and secretive backdrop enabling refuge or recovery, a place of transition, a pivot point within a story.

It is from the cave on the deserted island of Monte Cristo that Edmund Dantès finds the treasure which allows him to exact his spiralling plan of revenge in Alexandre Dumas's epic novel. Likewise, John Trenchard, the hero of J. Meade Falkner's book *Moonfleet*, evades capture from the excise men and recuperates in a sea cave, 'some eight yards square and three in height', where 'when the wind blows fresh, each roller smites the cliff like a thunder-clap'.

Cave-lore, I had come to discover, was also abundantly associated with similar, real-life stories. In the forest of Rothiemurchus, there is the Cat's Den – once the crag-top cave of a villainous local fugitive named Black Sandy, which several years previously I had set out to find. Various folk tales intersected around this secret spot, and I had triangulated them to identify the cave's rough location. It was, it turned out, the perfect spot for an outlaw to hide: small but perfectly habitable for one person; hidden from view, but also commanding a wide panorama above the blue-green canopy of ancient pines.

There is also Balnamoon's Cave, an incredibly well-concealed boulder den in the remote upper reaches of Glen Mark, where James Carnegy, the 6th Earl of Balnamoon, hid following the Battle of Culloden. For a year the Jacobite laird managed to evade capture, running to earth whenever the Redcoats closed in on him. Eventually, though, he was caught; a Presbyterian minister informed the government forces of his hiding place, which otherwise would have remained undiscovered.

Centuries may pass, but the mark of such events on these hidden places is so generationally memorised as to be indelible. On a slope of coastal farmland on the Rhins of Galloway, a narrow space between a cluster of moss-covered granite blocks reveals a small inner chamber. Local knowledge still claims this to be the cave of Sir Andrew Agnew, a recusant baronet 'driven to . . . seek shelter in strange hiding places'. Agnew, like so many other Presbyterians in that region, was hounded from his home by the feared Highland Host, a militia given licence to plunder the possessions of suspected Covenanters. He found refuge in a cave with a 'small aperture through the heathery hill' described by a descendant of his as being 'the haunt of the sea-otter or the chough'.

Caves, like other discreet landscape features, have also been venerated for spiritual significance or contemplative value. Again, there is the feeling of claim and concealment, of places appropriated, made quietly special. St Ninian is reputed to have occupied a sea cave near Whithorn in Galloway, giving rise to centuries of devotional journeying by pilgrims to the site. Millennia-old crosses carved into the cave's coarse sandstone describe the strength of this connection, bearing witness to its saintly connection.

Elsewhere, caves have been subject to other consecrating acts. Within a sea cavern on the rocky tidal island of Davaar, off the Kintyre Peninsula, there exists a life-size painting of Christ on the cross. In 1887 an art teacher, Archibald MacKinnon, experienced a vision compelling him to depict the Crucifixion within a specific cave on the island. He travelled there in secret, painting directly onto the cave wall and telling no one of his mysterious undertaking. When the image was eventually discovered by local fishermen, it was thought to be a miracle – the result of divine, rather than human, placement. However, as the story of its creation was finally revealed, the local population are said to have been angered at what they must have taken to be an elaborate dupe, ostracising MacKinnon until he left his home, never to live in the area again.

There is evidence too of much older ritualistic practices. In 1928 the archaeologist Sylvia Benton began excavating what has become known as the Sculptor's Cave at Covesea on the southern coast of the Moray Firth. Her discoveries within the barely accessible sea-cliffs were astonishing in their scale. Alongside Late Bronze Age metalwork and Roman Iron Age artefacts, a huge concentration of human remains were found. The bones, a significant number of which were from children, spanned a multi-century time period and showed signs of disarticulation as well as deliberate arrangement, implying a site of a long-standing funerary prominence. Incised on the entrance walls of the huge mortuary complex, a collection of Pictish symbols was also identified – crescent shapes, pentacles, a triple oval – dating from much later, around 600–800 AD. The petroglyphs are indecipherable but undoubtedly coincide with somewhere of deep ceremonial importance: a liminal place at the dynamic edge of sea and land, the transitional zone between one world and the next.

But knowledge of the history of Scotland's caves cautions against benign sentimentality. On the Hebridean island of Eigg, a small entrance point fissured between damp, algae-covered rocks leads to Uamh Fhraing – once the site of a notoriously brutal act of clan warfare. In the winter of 1577, the island's inhabitants, the MacDonalds of Clanranald, spotted a raiding party of their avowed enemy, the MacLeods, sailing over from the Isle of Skye. Following years of bloody clan feuding, the islanders quickly understood the threat they were facing and hid within the confines of the cave.

When the MacLeods landed they found Eigg seemingly deserted. For three days they searched for the Macdonalds, looking to avenge an attack that had been made on their laird's son. Their hunt, however, proved fruitless. The MacDonalds were nowhere to be seen, and with heavy snow covering any sign of tracks, the MacLeods decided to leave.

But as they sailed away, the raiders spotted the unmistakable form of a lone human figure. The MacLeods doubled back, finding fresh prints in the snow, which they then followed to the cave. Immediately, the islanders' means of concealment became the instrument of their containment. A huge fire was lit at the cave's entrance, smoke billowing through its tiny cavity and suffocating everyone inside. Nearly the entire population of Eigg are believed to have been murdered in Uamh Fhraing, almost 400 people meeting their death in what has been referred to ever since as the Massacre Cave.

As I glanced upwards, the hillside suddenly came alive, a large swathe of the slope abruptly moving sideways. The land appeared to rise, flowing westwards, rippling like a sheet lifted from a bed. For a moment I struggled to focus, my eyes still sleep-lazy in the flat, early morning light.

It was the day after the rainless storm and, as the winds receded, a succession of petulant squalls had followed in its wake. The blue sky had gone too, replaced by flinty clouds and air heavy with moisture. We'd returned to Inchnadamph just after dawn, this time walking the route of a different burn four kilometres further south, the Allt nan Uamh. As the stream met the River Loanan, we struck out on a well-used path, past a static caravan raised above the ground on struts, its grey curtains drawn tightly shut against the morning.

We had entered into another narrow-breached valley, stepping through a causeway of rutted limestone, when I saw the deer. The herd was enormous: two hundred or more, a tightly packed mass of frightened bodies, startled into swift, unified motion. Seconds earlier, standing still against the heather half a kilometre away, they had been entirely camouflaged. Now, though, making a curving run, up and then down the flank, it was as if the ridge itself was moving, the ground loosened, turning to liquid.

For a few minutes afterwards, we talked at double speed, my son's face bright with wonder at what he'd just seen. Still distracted, we forded a wide burn that ran across our path, not realising its strangeness. The water was moving with force, bubbling as it flushed around boulders, breaking white at our ankles. Below the surface I could see the bright green fronds of waterweed trailing in the current. Shallower water pooled calmly above, but then, nothing. The burn's upward reach immediately ended, blocked by the rising bank of hillside. I looked higher, expecting to see a hidden stream, a downward flow from somewhere, but the ground was dry. Then and there, the stream simply came into being, upwelling through shale and rocks, a fully formed burn gushing into life. I thought about the disappearing waterway we had seen the day before and the roaring, subterranean cascade in Uamh an Tartair. This was the counterpart mystery, the revelation of the landscape's sleight of hand.

There were other unexpected waypoints. On the banks of the Allt nan Uamh we found sinkholes hollowed in the grass, some six feet wide and almost a foot deep. Inside each one, settled on a bed of coarse sand, were clutches of large, smoothed-sided boulders deposited and shaped by the overflow of water so that they resembled enormous eggs, and the craters themselves, the nests of some fantastical creature.

As we crossed the Allt nan Uamh's dry riverbed, I paused in the middle, turned sideways into what would be the river's flow and looked upstream. A road of grey-white moraine chicaned higher into the valley before disappearing around a tight bend. The scene felt surreal – desolate and ghostly. Never before in Scotland had I come across a completely arid waterway. I wondered if this was more of the landscape's trickery, a once ancient system diverted and drained elsewhere, or the result of something more desperate, more globally devastating.

The path began to rise steeply as we headed towards a large limestone outcrop, jutting from the ridge in a series of metallic-grey ramparts. At the base of the first large rock face and above a talus of rock that fanned sharply down into the glen below we reached a row of dark archways, half-moon openings embedded in the cliff. These were the Bone Caves, so called because of the extraordinary archive of animal remains that over the years have been discovered in them.

Peach and Horne first excavated the caves in 1889, finding brown bear and reindeer bones, while later digs unearthed a wide array of other extinct fauna, from lemming, lynx and arctic fox to wolf and wild horse. The quantity and diversity of the finds have rendered the caves with a reliquary significance that is unparalleled anywhere in Scotland and which is not readily explained. Although there is evidence of human presence, including Neolithic interment and most unusually a pin made from walrus ivory, many of the animal bones predate the oldest human relics by thousands of years, so were not brought there by hunters. Instead, the caves shed light on Scotland's natural history during the period between the retreating ice sheets, when post-glacial conditions supported a much wider range of boreal wildlife. The accumulation of animal bones may be little more than the effect of time and environment: the remains of creatures that found their way into the caves over millennia, or fragments washed in with the melting ice.

As we neared the entrance of the first cave, we were greeted by a voice from the slope below. A man dressed in a down jacket and wearing winter mountaineering boots approached us, picking his way nimbly up through the moraine. 'I see you've found my home,' he said, smiling through a thick, sandy beard. 'Come on, I'll show you round.'

We followed him under a wide overhang at the base of the cliff. A roll mat and sleeping bag had been laid out on the cave's floor. Next to them was a torch and a book whose pages were curling at the corners. On a ledge of rock to the side, a makeshift cooking area had been fashioned: a collapsible stove and blue gas canister sat between pans and bags of sealed food. 'This place has been a godsend,' he explained, telling us how he had been traversing the peaks of Conival and Ben More Assynt two days previously when the gales had unexpectedly rolled in. 'I'd been planning to sleep on the tops,' he continued, gesturing towards the hills on the opposite side of the glen, 'but the conditions were unbearable. I spotted the caves from the ridge and thought they'd be the perfect place to sit out the storm.'

I liked that the hillwalker had found the caves; had needed and sought their protection. There was something decidedly uncomplicated in the instinct, an age-old act that resonated strongly with me. Besides, the place was an incredible location to overnight. It was dry and as spacious as a child's bedroom. The ground was mainly earth, packed hard and level enough in the middle to sleep comfortably on. From inside there was an elevated view through the entrance. Shafts of steep hillsides, sky pitched above, the stone river flowing, but not flowing in the glen below. It remined me of the Priest's Hole cave in Cumbria, another penthouse howff which possessed that rarest of combinations: shelter and height.

We explored inside. At the rear of the cave, the floor sloped away to a smaller separate level, big enough to retreat to in the worst of weathers. Within the walls we found cracks and fissures, darker spaces glimpsed beyond. In one section, the rock had dissolved in segments, leaving cavities in the wall that formed shelves and cubby holes no bigger than a fist. As we investigated the other caves, I sensed a connection between them all, each a discrete place linked somehow, a network of channels, chambers and conduits that had spread like roots through the rock.

Stepping back outside, the landscape seemed suddenly more permeable. Stream cuts wrought lines across the hillsides; flanks and ridges appeared gently moulded, smoothed and folded into one another; frost-shattered quartzite lay winnowed on the peaks; and as the sun pressed behind the clouds, it gave through in a soft, watery light.

Wood of the Ancient Well

Munlochy, the Black Isle

We've pulled off the road and onto a gravel track tucked among some tall pines. After a couple of false starts, eyeing-up other stands of woodland in this part of the Black Isle, I'm sure this must be the place. The track opens into a wide area, big enough to take a dozen vehicles or more. It tells of somewhere significant, regularly visited. There's another car there already, three people loitering by the boot eating sandwiches. They're all adults, but from the ages of the oldest two and the younger woman's body language, I can tell they are a family. We park a little distance from them and for some reason wait until they've moved on before we get out.

A fine drizzle sifts down through the trees. There's the washed-clean smell of pine needles and the aroma of tree sap. Somewhere deeper in the wood I can hear bird call, faint and melodious. My wife puts on her waterproof and ties a scarf. I look for something to indicate where we should go but see nothing except a narrow path cleaved between some bracken. It's as good a place as any to start, and we cut through the ferns uncertain in our direction until up ahead, we start to see them. My wife is in front of me and points to them first, a hundred metres away, just off the track. They're hung at head height from the branches of ash and birch saplings. The colours are multiple, but my eyes register mainly whites and greys. Even at a distance, they appear lurid and out of place, clashing with the background green of the forest, shapes that lack the intermeshing lock of foliage; something unnatural – not of the wood.

The objects resolve themselves as we get closer, turning into recognisable items. There are socks, scarves, an odd glove and a hankie. I can make out the remains of t-shirts, pyjamas and even a pair of trousers tied by their legs round the trunk of a tree. Mostly though, it is just material: ribbons, rags, patterned prints, fabrics felt-tipped with dates and people's names. I study it all with a detached fascination until I see a baby's bib attached by its clasp around a thin branch. And not for the first time on my journeys I'm stung with a sudden sense of poignancy, a moment of something heartfelt for the unknown person who left it there.

As we continue, the offerings multiply, weighing on every tree. They begin to take different and stranger forms. Plastic bags, flags, an umbrella and most bizarrely a power tool dangling by its electric cord. The path rises up a bank, and we follow it to its crest. As I catch a first glimpse down onto the other side, the sight pulls the air from my lungs. Below us the entire wood is bedecked in thousands of placements. Every available branch seems to have been covered – knotted and down-hanging with some form of drapery. The effect is startling, creating the look of somewhere disaster-impacted: the aftermath of a flood, the wake of a typhoon. But it's not. It's something else, something deliberate and immeasurably meaningful. I can see that the scene converges, centres on a narrow gulch between the trees where a cleft in the ground reveals the site of an ancient well.

The spring is dry, but a channel is worn in the earth, leading back to a hole hollowed from the bank like a badger sett. More rags are tied to roots around the opening. Many are old, bleached of their original colour, stained green and disintegrating. This is part of the tradition. The rags are called cloots, pieces of cloth that are to be dipped in the waters of the well and held against parts of the body affected by disease or ailment. As the cloots are hung and decay over time, so the patient is supposedly simultaneously relieved of their sickness. The cloots are also indicative, votive offerings made by loved ones that belong to a person for whom a prayer of supplication is being made.

Places such as this are compelling for the ritual devotion they summon, a repeated association of people to a specific point in the landscape. Often these relationships are enduring beyond individual lifespans, shared communally through the ages, their significance passed on over multigenerational time-periods. 'We create our meanings from rocks and air and fire and water; and sacred springs and holy wells have always been considered, from their earliest days, to be places where the meanings of this world fade and other realities are revealed,' observed Phil Cope, in his extensive explorations of Scotland's wellsprings. They are, he continues, 'where people for millennia have journeyed to . . . leave their own small markings of connectedness with lives lived and beliefs once embraced'.

This seemed true for many of the places I had visited on my journeys. The ruined chapel on St Finnan's Isle, the standing stone at Camas nan Geall and the burial ground on Inishail had suggested a sense of continued reverence, of sanctity preserved and deepened. To consider the weight of this spiritual antiquity is to realise that there are wild landscapes which transcend our modern notions of utility and purpose, that there are places whose importance may never be fully understood yet which continue to resound with mystery and meaning.

There is a cultural resonance associated with them too. A soft reverberation still pulsing in our imaginative perceptions of Scotland's wild places. It can be heard echoing in the literature of the open road or caught in the plaintive words of nostalgia for lost or forsaken homesteads. It is there enigmatically, but no less powerfully, in the oral traditions. Carried on the bindweed tendrils of folklore, or shared selectively and secretly – a whisper in an ear, hushed tones spoken to describe a hidden place: an illicit still, an osprey's nest, an undisclosed cave perhaps.

Scotland's wild histories, I had come to realise, are the origins of legacies as profound as they are unfamed. The sinuous networking of drovers' routes, the wild enclaves of the Christian missionaries, the emptiness of Clearance villages and the ghostscapes of industrial decline tell of more than their mere physical abandonment. They speak of immeasurable social change, of the movement of people through time and the landscape. Yet they are often lost, overlooked, or passed by. Our attention is diverted somehow from their significance, as if remoteness casts them with muted relevance or immediacy. The opposite, in fact, is true. For wildness is part of their story, then and now, the common denominator defining and shaping their narratives.

To find these places, therefore, requires little more than a resetting of focus. An adjustment in our sightlines, both literally and figuratively. That, and the obligation, as the Scottish nature writer Jim Crumley puts it, 'to walk the landscape and see what rubs off'. Only by doing so are we able to more closely interpret the complex interactions of past lives within a certain space. In this way, our personal responses to wild histories become vital. To fully understand 'the poetics and politics of paths and places', the archaeologist Christopher Tilley asserts that we must master 'the art of walking in and through them, to touch and be touched by them'. Tilley's radical proposition, that 'knowledge of landscapes, either past or present is gained through perceptual experience', seems particularly relevant for Scotland, where the weight of human history is so layered, yet often so forgotten in the landscape. The approach of phenomenology that Tilley pioneered as an archaeological discipline is both intuitive and democratising, empowering any individual with the ability to determine the meaning of histories and wildness just by getting out there. 'For the phenomenologist, his or her body is the primary research tool,' Tilley explains. 'He or she enters into the landscape and allows it to have its own effect on his or her perceptive understandings . . . This approach', he says, 'means accepting that there is a dialogic relationship between person and landscape.'

This sense of correspondence with landscape, the consensual traffic between person and place, had been with me throughout my journeys. I had felt it in the imaginative projections of the Glen Loin Caves, in the retracing of Jock's Road and as I surveyed the once seething industrial panorama of the Blackwater Reservoir. It had been there too in the visceral immediacy of Bourblaige, Belnahua and the Atlantic Wall, as well as the many memorials and human mementos I had come across. At times it had felt impossible, even wrong to look for a rational, objective response. Instead, it seemed simpler to defer to instinctive emotions: the tranquillising calmness that came from the time spent on Loch Awe's islands, the night in Davy's Bourach, and my crag-top evening view across to the Slate Isles. Or, by contrast, the distinctly unsettling atmosphere of the Loch Chiarain Bothy and Inchkeith.

It was, though, this intuitive reaction which had so intrigued me. During all of my journeys I had come to think that wild histories seemed somehow indefinably linked with the spirit of a place, overlapping, meshed and entwined. That the 'subtle marks' which the writer Kathleen Jamie references were indeed part of the *genius loci* of certain places. There as part of (or because of) the energy that some landscapes hold.

Talking about the connection he has sensed with primary landscapes in his native Australia, the novelist Tim Winton I think puts it best: 'Is this resonance just a signal of the life force in the country?' he asks. 'In spots like these it can be a relief to find evidence of ancient culture. The petroglyph, the rubbing stone or ochre painting lets you off the hook. You can reassure yourself that someone else has felt this before you. So perhaps you are not imagining it. But then you wonder: am I feeling the people of this place or the power they have always found in it?'

We step slowly between the trees, the ground beneath us bronze with fallen beech leaves. Above us there remains a ragged dome of vaulted green light. Somewhere in the distance I can hear the sounds of a chainsaw and the muffled tones of a shouted conversation. But inside the wood it feels calm, a contained peacefulness that seems close to being churchly. All the time I'm trying to understand what's in front of me, but I can't quite manage, dumbfounded by the scale of it all. The mass of rags is too much to take in, blurring to a single, muted colour. And then it suddenly hits me, the hugeness of what I'm seeing: a profound spectacle of human emotion, all the hope and the fear and the love that so many people have left at this place.

We stay there for a while in that halted, meditative state before cresting the small rise of trees, ducking between branches and the fall of offerings, where we see people walking in the wood below us, each some distance apart, treading quietly to the same spot.

Acknowledgements

I'd like to thank the people that have joined me at different stages on the journeys researching this book. Specifically, Chris Burton for his unfailing appetite for unusual venturing. Steve Owen for bringing practical help and his usual steadfast support and humour while exploring the islands of Loch Awe. Alice McInnes (seakayakalice.co.uk), for her good company and guidance in the waters of the Forth. Tony Hammock (seafreedomkayak.co.uk), whose unbridled enthusiasm for, and deep knowledge of, the Firth of Lorn made for a fascinating trip. Although not directly involved in the journeys for this particular book, I would like to thank Simon Gall and Ben Kellett for generously sharing their skills and experience in the outdoors with me over the years. I'd also like to thank the following people who provided information that was used as part of the book's research: Jonathan Haylett; Jess Dinning at the Dalmally Historical Association; Fiona Holmes at the Northern Lighthouse Board; Lynda McGuigan at the Museum of Scottish Lighthouses; Dr Laura Hamlet at NW Highlands Geopark; and Sue Agnew at Scottish Natural Heritage. My appreciation goes to Neil Evans for his careful proofing of the final manuscript, Helen Bleck for her considered copy-edit, and Andrew Simmons and the staff of Birlinn for the production of the book. Thanks too to Alan McKirdy, who kindly provided a sense check of the geological information mentioned in the text. I am also indebted to my parents, Angela and Melvyn Baker, as well as John and Noreen Young, for all their powerful encouragement. Finally, I would like to express my thanks and my love to my wife Jacqui and my children, Isla, Rory and Finn, for their unfaltering patience in the years I was researching and writing this book.

Author's note: leave no trace. Throughout the journeys made for the book, I wanted to minimise any intrusion on the landscapes I was travelling in. After bivvying and camping, each spot was cleared of any imprints and checked for even the smallest fragments of debris. Although in one of the chapters I have described the use of a small campfire, this was lit in a steel pan where there was absolutely no risk of the fire spreading either above or below ground and where no scorch marks would remain. Fuel for the fire was also brought with me rather than taken from the location, ash and charred wood was collected and taken back with me, and the site was brushed clean. Likewise, every effort was made not to disturb the historic objects or constructions I came across.

Bibliography

Agnew, Sir Andrew. *A History of the Hereditary Sheriffs of Galloway*. A & C Black, 1864.

Barkham, Patrick. *Islander: A Journey around Our Archipelago*. Granta, 2017.

Barnes, Patricia M. and Barrow, G.W.S. 'The Movements of Robert Bruce between September 1307 and May 1308'. *The Scottish Historical Review*, Vol. 49, No. 147, Part 1 (1970), pp. 46–59.

Bathhurst, Bella. *The Lighthouse Stevensons*. HarperCollins, 2005.

Beckett, Samuel. *Waiting for Godot*. Faber & Faber, 2006 (reprint).

Borthwick, Alastair. *Always a Little Further*. Diadem, 1983.

Brown, Hamish. *Hamish's Mountain Walk*. Sandstone Press, 2010.

Brown, Philip. *The Scottish Ospreys: From Extinction to Survival*. William Heinemann Ltd, 1979.

Bunting, Madeleine. *Love of Country: A Hebridean Journey*. Granta, 2016.

Cameron, Donald. *The Field of Sighing: A Highland Childhood*. Birlinn, 2003.

Campbell, Alexander. *The History of Leith, From the Earliest Accounts to the Present Period*. William Reid & Sons, 1827.

Connor, Jeff. *Creagh Dhu Climber: The Life and Times of John Cunningham*. Ernest Press, 1999.
Cope, Julian. *The Modern Antiquarian: A Pre-millennial Odyssey through Megalithic Britain*. Thorsons, 2011.
Cope, Phil. *Holy Wells Scotland*. Seren, 2015.
Cowen, Rob. *Common Ground*. Windmill Books, 2016.
Crumley, Jim. *The Last Wolf*. Birlinn, 2015.

Davies, W.H. 'The Forsaken Dead' in *New Poems*. Elkin Mathews, 1907.
Davies, W.H. *Autobiography of a Super-Tramp*. Parthian Books, new edition 2013.
Deakin, Roger. *Waterlog*. Vintage, 2000.
Defoe, Daniel. *A Tour Through the Whole Island of Great Britain*. Penguin Classics, reprinted 1978.
Devine, T.M. 'The Rise and Fall of Illicit Whisky-Making in Northern Scotland, c. 1780–1840', *The Scottish Historical Review*, Vol. 54, No. 158, Part 2 (1975), pp. 155–77.
Devine, T.M. *Clanship to Crofters' War: The Social Transformation of the Scottish Highlands*. Manchester University Press, 1994.
Devine, T.M. *The Scottish Clearances: A History of the Dispossessed, 1600–1900*. Penguin, 2019.
Dickson, John. *Emeralds Chased in Gold; or, the Islands of the Forth: Their Story, Ancient and Modern*. British Library, Historical Print Editions, 2011.
Donaldson, M.E.M. *Wanderings in the Western Highlands and Islands*. Alexander Gardner, 1923.
Dumas, Alexandre. *The Count of Monte Cristo*. Penguin Books, new edition 2003.

Elphinstone, John. *A New and Correct Map of North Britain*. And. Millar, 1745.

Fawcett, Percy. *Exploration Fawcett*. Weidenfeld & Nicolson, 2010.

Gillen, Con. *Geology and Landscapes of Scotland*. Dunedin Academic Press, 2nd revised edition 2013.

Goldsmith, Oliver. 'The Deserted Village'. W. Griffin, 1770.

Goodenough, Kathryn M. & Krabbendam, Maarten. *A Geological Excursion Guide to the North-West Highlands of Scotland*. NMSE Publishing Ltd, 2011.

Gordon, Seton. *Days with the Golden Eagle*. Whittles Publishing, new edition 2003.

Graham, Stephen. *The Gentle Art of Tramping*. Budge Press, 2007.

Grant, James. *Old and New Edinburgh*, Vol. VI. Cassell & Company Limited, 1880.

Grjotheim, Kai & Kvande, Halvor (eds). *Introduction to Aluminium Electrolysis:Understanding the Hall-Héroult Process*. Beuth Verlag, 2011.

Gunn, Neil M. *Highland River*. Arrow Books. 1975.

Hagan, John. 'A Note on the Significance of Diggory Venn', *Nineteenth-Century Fiction*, Vol. 16, No. 2, pp. 147–55. University of California Press, 1961.

Haldane, A.R.B. *The Drove Roads of Scotland*. Birlinn, new edition 2015.

Hamerton, Philip Gilbert. *An Autobiography 1834–1858*. Seeley & Co. Limited, 1897.

Handley, James Edmund. *The Navvy in Scotland*. Cork University Press, 1970.

Hardy, Thomas. *The Return of the Native*. Wordsworth Editions, new edition 1995.
Harris Dick Ogilvy, Eliza Ann. *A Book of Highland Minstrelsy*. Kessinger Publishing, new edition 2008.
Hart-Davis, D. *Monarchs of the Glen: A History of Deer-Stalking in the Scottish Highlands*. Jonathan Cape, 1978.
Haswell-Smith, Hamish. *An Island Odyssey*. Canongate, new edition 2014.
Hendrie, William F. *The Forth at War*. Birlinn, 2005.
Hill, Peter. *Stargazing: Memoirs of a Young Lighthouse Keeper*. Canongate, 2003.
Hobsbawm, E.J. 'The Tramping Artisan', *Economic History Review*. Series 2, III (1950–1), p. 313.
Hunter, James. *On the Other Side of Sorrow*. Birlinn, 2014.

Jamie, Kathleen. *Findings*. Sort of Books, 2015.
Johnston, Thomas. *The History of the Working Classes in Scotland*. Forward Publishing, 1920.
Jones, John Paul. *Memoirs of Rear-Admiral Paul Jones now first compiled from his original journals and correspondence*. Oliver & Boyd, 1830.

Kempe, Nick & Wrightman, Mark (eds). *Hostile Habitats: Scotland's Mountain Environment*. Scottish Mountaineering Trust, 2006.
Krauskopf, Sharma. *Scottish Lighthouses*. Appletree, 2001.

Lawrence, D.H. 'The Man Who Loved Islands', in *Selected Stories*. Penguin Classics, 2007.
Lawrence, Martin. *The Yachtsman's Pilot: Skye and North-west Scotland*. Imray Laurie Norie & Wilson Ltd, 2010.

Lee, Laurie. *Red Sky at Sunrise: Cider with Rosie, As I Walked Out One Midsummer Morning, A Moment of War.* Penguin, new edition, 2014.

London, Jack. *The People of the Abyss*. CreateSpace Independent Publishing, 2013.

Lucas, E.V. *The Open Road*. Methuen & Co, 1931.

Lusby, Philip and Wright, Jenny. *Scottish Wild Plants*. Royal Botanic Garden Edinburgh, 1996.

MacCaig, Norman. 'The Pass of the Roaring', written 1972, published in *The World's Room*. Chatto & Windus, 1974.

MacCaig, Norman. *Between Mountain and Sea: Poems From Assynt*. Polygon, 2018.

MacFarlane, Robert. *The Wild Places*. Granta Books, 2007.

MacGill, Patrick. *Children of the Dead End*. Birlinn Ltd, new edition 1999.

MacGill, Patrick. *The Rat-Pit*. HardPress Publishing, 2013.

MacGregor, Alasdair Alpin. *The Peat-Fire Flame*. The Moray Press, 1937.

MacLean, Sorley. 'The Tree of Strings', in Emma Dymock & Christopher Whyte (eds), *A White Leaping Flame/ Caoir Gheal Leumraich Sorley Maclean: Collected Poems*. Polygon, 2011.

MacLeod Barron, Evan. *The Scottish War of Independence: A Critical Study*. R. Carruthers & Sons, 2nd edition 1934.

McGonagall, William. 'The Tragic Death of the Rev. A.H. Mackonochie', 1887.

McKirdy, Alan. *Set in Stone: The Geology and Landscapes of Scotland*. Birlinn, 2015.

McKirdy, Alan; Gordon, John & Crofts, Roger. *Land of*

Mountain and Flood: The Geology and Landforms of Scotland. Birlinn, 2017.

McOwan, Rennie. 'The Glen Loin Cavemen'. *The Scots Magazine*, 144 (3) (1996), pp. 263–8.

Meade Falkner, John. *Moonfleet*. Puffin Classics, new edition 2018.

Milliken, William & Bridgewater, Sam. *Flora Celtica*. Birlinn, 2004.

Moir, D.G., Bennet, D.J. & Stone C.D. (eds). *Scottish Hill Tracks*, 4th edition. Scottish Mountaineering Trust, 2004.

Morris, R. & Barclay, G. 'The fixed defences of the Forth in the Revolutionary and Napoleonic Wars, 1779–1815'. *Tayside and Fife Archaeological Journal*, Vol. 23 (2017), pp. 109–33.

Morrison, Ian. *Landscapes with Lake Dwellings. The Crannogs of Scotland*. Edinburgh University Press, 1985.

Morrison-Low, Alison. *Northern Lights: The Age of Scottish Lighthouses*. NMSE Publishing, 2010.

Muir, Richard. *The Lost Villages of Britain*. Michael Joseph, 1982.

Nancollas, Tom. *Seashaken Houses: A Lighthouse History from Eddystone to Fastnet*. Particular Books, 2018.

National Records of Scotland. 'Macpherson v Scottish Rights of Way & Recreation Society Ltd', 1887. 14 R 875 and (1888) 15 R (HL) 68.

Oban Times. 'Island of Easdale Submerged', 3 December 1881.

Ochota, Mary-Ann. *Hidden Histories*. Francis Lincoln Ltd, 2016.

Oldroyd, David R. *The Highlands Controversy: Constructing Geological Knowledge Through Fieldwork in Nineteenth-Century Britain*. University of Chicago Press, 1990.

Orwell, George. *Down and Out in Paris and London*. Penguin Classics, new edition 2001.

Peach, B.N., Horne, J., Gunn, W., Clough, C.T., Geikie, A., Hinxman, L.W. & Teall, J.J.H. *The Geological Structure of the North-West Highlands of Scotland*. HMSO, 1907.

Pennant, Thomas. *A Tour in Scotland 1769*. Birlinn, new edition 2000.

Perchard, Andrew. *Aluminiumville: Government, Global Business and the Scottish Highlands*. Crucible Books, 2012.

Pollard, Tony & Banks, Iain. 'Archaeological Investigation of Military Sites on Inchkeith Island', in Pollard & Banks (eds), *Bastions and Barbed Wire: Studies in the Archaeology of Conflict*. Brill, 2009.

Prebble, John. *The Highland Clearances*. Penguin, new edition, 1982.

Pryor, Francis. *The Making of the British Landscape*. Penguin, 2010.

Rackman, Oliver. *The History of the Countryside*. J.M. Dent & Sons Ltd, 1986.

Ransome, Arthur. *Swallows and Amazons*. Vintage Children's Classics, new edition 2012.

Reynolds, Michael. *Martyr of Ritualism: Father Mackonochie of St. Alban's, Holborn*. Faber & Faber, 1965.

Saville, Alan. 'Archaeology and the Creag nan Uamh bone caves, Assynt, Highland'. *Proceedings of the Society of Antiquities of Scotland,* 135 (2005), pp. 343–69.

Schama, Simon. *A History of Britain.* BBC Worldwide Ltd, 2000.

Schama, Simon. *Landscape and Memory.* HarperCollins, 1995.

Scott, Sir Walter. 'Marmion'. Archibald Constable (Edinburgh) and John Murray (London), 1808.

Scott, Sir Walter. *Chronicles of the Canongate: The Two Drovers.* Penguin Classics, new edition 2003.

Sinclair, Sir John (ed.). *Old Statistical Account of Scotland.* 1791–9.

Smout, T.C. & Stewart, Mairi. *The Firth of Forth: An Environmental History.* Birlinn Ltd, 2012.

St John, Charles. *A Tour in Sutherland with Extracts from the Field-Books of a Sportsman and Naturalist.* John Murray, 1849.

St John, Charles. *The Wild Sports & Natural History of the Highlands.* Macdonald Futura Publishing Limited, 1981.

Stephanides, Stephanos & Bassnett, Susan. 'Islands, Literature, and Cultural Translatability'. *Journal of Global Cultural Studies* 2008, pp. 5–21. http://journals.openedition.org/transtexts/212.

Stevenson, Robert Louis. 'The Vagabond', in *Songs of Travel and Other Verses.* Chatto & Windus, 1896.

Stevenson, Robert Louis. *Treasure Island.* Modern Library, new edition 2001.

Strachan, Michael. *Scottish Lighthouses: An Illustrated History.* Amberley Publishing, 2016.

Thompson, E.P. *The Making of the English Working Class*. Victor Gollancz Ltd, 1963.

Thomson, Ian. *May the Fire Always be Lit. A Biography of Jock Nimlin*. Ernest Press, 1996.

Thomson, Ian. *The Black Cloud: Scottish Mountain Misadventures, 1928–1966*. Ernest Press, 1993.

Tilley, Christopher. *Body and Image: Explorations in Landscape Phenomenology 2*. Routledge, 2008.

Tilley, Christopher. *Interpreting Landscapes: Geologies, Topographies, Identities; Explorations in Landscape Phenomenology 3*. Routledge, 2012.

Towle, E.A. & Russell, E.F. (eds). *Alexander Mackonochie: A Memoir*. Kegan Paul, Trench, Trübner, & Co. Ltd, 1890.

Wightman, Andy, Higgins, Peter, Jarvie, Grant & Nicol, Robbie. 'The Cultural Politics of Hunting: Sporting Estates and Recreational Land Use in the Highlands and Islands of Scotland'. *Culture, Sport, Society*, Vol. 5, No. 1 (2002), pp. 53–70.

Wightman, Andy. *The Poor Had No Lawyers: Who Owns Scotland and How They Got it*. Birlinn Ltd, 2015.

Wightman, Andy. *Who Owns Scotland*. Canongate, 1996.

Winton, Tim. *Island Home: A Landscape Memoir*. Picador, 2016.

Withall, Mary. *The Islands that Roofed the World: Easdale, Seil, Luing and Belnahua*. Luath Press, 2001.

Wood, Emma. *The Hydro Boys*. Luath Press, 2004.

Index

Aberdeen, Lord 57
Aberdeenshire 98, 111
Aberdour 160
Adomnán, St 168
Agnew, Sir Andrew 189
Ailsa Craig 104
Allt an Loch 47
Allt nan Uamh 192, 193
Amazon, River 107
Anstruther 160
Ardnamurchan 85, 93, 94, 104, 105, 114–16
Arisaig 85
Arran 104
Arrochar Alps 4
Association for the Protection of Public Rights of Roadway 50
Assynt 177, 179, 180, 181, 184
Atholl 50
Atholl, Duke of 201
Atlantic Wall, Sheriffmuir 91, 92

Baldred of Tyninghame, St 168
Ballachulish 82, 119, 128
Ballachulish, North 30
Balnamoon's Cave, Glen Mark 189
Bass Rock 168, 170
Bay of Skaill 110
Beckett, Samuel 19
Bell Rock Lighthouse 135, 136
Belnahua 7, 118–20, 126, 127, 130, 132, 139, 141, 143, 145, 201
Ben Hiant 103, 104, 113
Ben More Assynt 195
Benton, Sylvia 190
Black Islands 79
Black Isle, the 197
Black Sandy (local fugitive around Rothiemurchus) 189
Blackwater 7, 11, 13, 30, 45, 201
Bone Caves, Inchnadamph 7, 177, 194
Bonnie Prince Charlie 178
Bonser, K.J. 41
Borthwick, Alastair 5, 9, 10, 16
Bourblaige 114, 115, 120, 201
Boyle, James 59, 60
Braemar 35, 46, 47, 53, 58
Brigadoon 111
British Home Fleet 174
Bronze Age 93, 191
Brown, Hamish 8
Brown, Philip 73, 74
Buachaille Etive Mor 16
Buckhaven 160
Bunting, Madeleine 114

Burntisland 161

Cairngorms 5, 15, 45
Callater Burn, near Braemar 35
Callater Stables 37
Campbell, Clan 116
Camas nan Geall 102, 115, 199
Canisp 179
Canna 92
Carn an t-Sagairt Mor 46
Carnegy, James 189
Cat's Den, Rothiemurchus 189
Caulfield, General 54
Celts, the 82, 93
Charles II 169
Chicken Rock Lighthouse 136
Choiremhuilinn 115
Ciaran mac an t-Saeir, St 106
Cladh Chiarain 106
Clunie Water 35
Clydebank 4
Cockenzie 159
Coire Breac 47, 58
Coll 92
Columba, St 85, 106
Conival 185, 195
Constable, John 78, 108, 109
Cope, Phil 199
Corryvreckan 145
Covenanters 169, 190
Covesea 190
Cowen, Rob 167
Craigleith 171
Crail 160
Creag Leachdach 52
Creag Phàdruig 46
Creagh Dhu 4
Crow Craigies 55, 60
Crumley, Jim 200
Cuan Sound 118
Cul Beg 179
Cul Mòr 179
Culloden, Battle of 88, 114, 189
Cumbria 150, 195

Daly, Frank 59, 61
Davaar Island 190
Davies, W.H. 20, 109
Davy's Bourach, Glen Doll 36, 57, 202
de Orellana, Francisco 107
Deakin, Roger 5
Defoe, Daniel 160
Devil's Beef Tub 15
Devil's Chair 15
Devil's Point 15
Devine, Tom 96, 97, 98
Devlin, Joseph 59, 60
Dubh Artach Lighthouse 137
Duffin, Harry 60

Dumas, Alexandre 188
Dun Mòr 121
Dunbar 149, 150, 152, 160, 162
Dunbar, William 70
Duncan, Rev. Grant 47
Dunwich 108, 109

Easdale 118, 119, 121, 122, 126–31, 140

Edinburgh 3, 50, 134, 147–51, 153, 154, 162, 166
Eigg 92, 191, 192
Eileach An Naoimh 145
Eilean Fhianain 64, 80, 83, 86, 87
Eilean Munde 82
Ellenabeich 120, 130, 131
Elphin 179
Elphinstone, John 39, 40

Fair Isle North Lighthouse 137
Fair Isle South Lighthouse 137, 138
Falls of Foyers 28
Fawcett, Percy 107
Fidra 171
Fife 151, 153, 159, 161
Finnan, St 64, 80, 85, 87, 107

Fisherrow 160, 162
Fladda 130, 132, 138
Flannan Isle Lighthouse 137
Firth of Forth 7, 134, 147–54, 160–64, 168, 171, 175, 205
Firth of Lorn 117, 130, 143

Galloway 189, 190
Garvellachs, 130
Geikie, Archibald 183, 184

Gleann Dubh 179
Glen Clunie 53
Glen Coe 119
Glen, Davy 57, 58, 60
Glen Doll 47, 50, 51, 54, 56, 58, 59
Glen Garry 69
Glen Leven 13, 14, 16, 32
Glen Loin 3–5, 187, 201
Glen Mark 189
Glen Tilt 40, 50
Goldsmith, Oliver 109
Gordon, Duke of 49
Gordon, Seton 69
Gosford Bay 161
Graham, Stephen 20
Grant, James 165
'The Grey Dog' (tidal reach between Lunga and Scarba) 145

The Haikes (fishing ground) 160
Hall, Charles Martin 28
Hamerton, Philip Gilbert 78, 79
Handa 82
Hardy, Thomas 19
'Harrying of the North' (genocide instigated by William the Conqueror) 110
Heroult, Paul Louis-Toussaint 28
'Highland Controversy' (geological disagreement) 180, 183, 184
Highland Host (militia) 189
Hill, Peter 138
Hobsbawm, E.J. 18
Horne, John 183–185, 194
Hound Point 161

Inchcailloch 82
Inchcape Rock 135
Inchcolm 168
Inchgarvie 152, 168
Inchkeith 148, 149, 151, 153–55, 159, 160, 166–68, 173–75, 202
Inchnadamph 180, 183, 192
Inishail 75, 78, 79, 82, 89, 199
Inner Hebrides 7, 92
Insh Island 129
Invercauld Estate 37
Invercauld, Lord 57
Iron Age 93, 108, 190
Isle Maree 82, 83
Isle of May 160, 168, 170

James IV 167
James VII 169
Jamie, Kathleen 202
Jock's Road (Glendoll to Braemar) 36, 47, 48, 51, 53–55, 58, 59, 61, 201
Jones, John Paul 149–52
Jura 145

Kinlochbeg 29
Kinlochleven 11–13, 23, 29, 30
Kinlochmore 29
Kintyre Peninsula 190
Kirkcaldy 151, 152
Knockan Crag 179, 181
Knoydart 69

Lairig Ghru 15
Lamb, the (island in the Forth) 171
Lammas Drave 161–64
Lapworth, Charles 181, 182
Largo 160
Lawrence, D.H. 167
Lee, Laurie 18
Leith 150, 151, 152, 153, 160, 166, 174
Lewisian gneiss 179
Lindsay, Robert of Pitscottie 167
Little Ross Lighthouse 137
Loanan, River 192
Loch Awe 64, 78, 202
Loch Callater 37
Loch Chiarain Bothy 33, 202
Loch Garten 73, 74
Loch Leven 29, 82
Loch Lomond 82
Loch Meadie 70
Loch Naver 70
Loch Phàdruig 46
Loch Shiel 64, 80, 85, 88, 90
London, Jack 20
Luing 118, 119, 126, 132, 138, 143
Lunga 145
Lyell, Charles 182

Mabey, Richard 158
MacCaig, Norman 61
MacDonalds of Clanranald 191
Macfarlane, Robert 167
MacGill, Patrick 11–14, 18, 21–24, 26
MacGregor, Alasdair Alpin 87

MacKinnon, Archibald 190
MacLeod, Clan 191, 192
Mackonochie, Rev. Alexander 32, 45
Macpherson, Duncan 50–52
Maol Rubha, St 82
Massacre Cave, Eigg 192
McFaul, Robert 59, 60
McOwan, Rennie 4, 5
Meade Falkner, J. 188
Meall Odhar 59
Methven, Battle of 3
Mesolithic 93
The Minch 179
Moidart 85, 86
Monro, Donald 119
Morar 85
Moray Firth 190
Mounth plateau 35
Muck 92
Muckle Flugga lighthouse 136
Muir, Richard 110
Mull 92, 104
Murchison, Sir Roderick Impey 180–83
Musselburgh 160

Napoleonic Wars 96, 153
Neolithic 93, 105, 111, 194
Nethersole-Thompson, Desmond 73

Newhaven 160
Nicol, James 181, 182
Ninian, St 190
Normandy 92
North Berwick 160
North Sea 153, 161

Ochil Hills 90
Oersted, Hans Christian 28
Ogilvy, Eliza Ann Harris Dick 81
Old Sarum 108, 109
Operation Osprey 74
Orkney 111
Orwell, George 19, 20, 145

Palaeogene 103
Pathhead Sands 151
Peach, Benjamin 183–85, 194
Peffer Sands 160
Pennant, Thomas 40, 119
Pentland Skerries 137
Pittenweem 160
Priest's Hole (cave in Cumbria) 196
Ptarmigan Club 4

Quinag 179

Raleigh, Sir Walter 107
Rannoch Moor 14, 16
Ransome, Arthur 63
Reformation 77
Rhins of Galloway 189
Riddell, Sir James 114–16
Robert the Bruce 3
Rockall 104
Ross of Mull 136
Rosyth 174
Rothiemurchus 188
Royal Society for the Protection of Birds 73

Rum 92, 104

St Finnan's Isle 81, 88, 89, 168
St John, Charles 70
St Kilda 104
St Monans 160
Sands of Forvie 111
Scarba 130
Scott, Sir Walter 43, 55, 136, 148
Scottish Borders 15
Scottish Ornithologist Club 73
Scottish Rights of Way and Recreation Society 50
Scourie 70
Sculptor's Cave, Covesea 190
Sedgwick, Adam 182
Seil Island 118–20, 126, 131
Sheriffmuir 90
Shetland 136
Shiel, River 92
Shirra, Rev. Robert 151
Sibbald, Sir Robert 161
Silurian Theory 181–83
Sinclair, Sir John 41–43
Skara Brae 111
Skerryvore Lighthouse 136
Skinnid 115
Skye 104, 191
Slate Isles 3, 118, 119, 202
Small Isles 92
Smith, Thomas 134, 135, 172
Smout, Christopher 162
Solway Firth 137
Sound of Luing 118, 126
Stac Pollaidh 179
Stevenson, Alan 136
Stevenson, David 136
Stevenson, Robert 135, 136
Stevenson, Robert Louis 20, 63
Stevenson, Thomas 136
Suilven 179
Sunart 85
Sutherland 70, 82

Thomson, Ian 60

Tilley, Christopher 200
Tiree 92, 136
Tolmount 59
Tornamona 115
Traith, the 160
Turner, J.M.W. 108, 109, 136

Uamh an Tartair 186, 193
Uamh Fhraing 191, 192

Vikings 83, 168

Walker, Ezekiel 135
Wars of Scottish Independence 168
Waterston, George 73
West Highlands 11, 86, 118, 184
West Lothian 161
White Water Burn 59
Whithorn 190
Wick 161
William the Conqueror 110
Winter Herrin' 161, 163, 164
Winter, John 57

Winton, Tim 202

Also available by Patrick Baker
The Cairngorms: A Secret History

'Perfect . . . full of the ghosts of walkers past'
Condé Nast Traveller

'Baker illuminates the bleak landscape'
FT Weekend

'Packed with great stories and vivid descriptions'
Scotland Outdoors

A secret history lies on the Cairngorm plateau – a series of barely known human and natural stories which have left few traces on the primordial landscape of Britain's last wilderness.

Patrick Baker has devoted years to exploring this history, searching remote mountain crags and discovering fascinating relics of the dramas of the past. He finds the skeletal remains of an aircraft, an ancient gem mine, an outlaw's hideaway and a mysterious aristocratic settlement. He wanders the hills in search of the legendary Big Grey Man of Ben Macdui, geological oddities and the elusive Scottish wildcat.

The result of his quest is a unique travelogue that explores the human impact of this stunning landscape, the story of the early days of mountaineering in the Cairngorms and the secrets of a past which is often stranger than fiction.

An excerpt from
The Cairngorms: A Secret History ('Ravine')

What was immediately obvious was that the Water of Ailnack was not immediately obvious. I had made my way south-west from the village of Tomintoul and was walking in the high moorland that flanked the river. Indigo clouds massed to the south, and a warm wind hinted at thunder. I passed a greenshank flying repetitive circuits, landing then returning to the air, its call a shrill *kew kew*.

I was less than 300 metres from the Water, yet it was impossible to see where it ran. It was there somewhere, embedded in the wide cleft that split the otherwise unbroken plateau of grass and heather. From further away even this feature would be lost. The landscape would appear continuous at eye level, the Water and its gully completely invisible.

On first impression this seemed like barren country. The distance between my feet and the horizon consisted of nothing but kilometres of tawny moor. And for a while I had the strange feeling of walking in a desert, somewhere featureless and immeasurable. Viewed carefully, though, the place took on a different perspective. With every few hundred metres I began to notice differences in the natural configurations of the moor. Heather gave way to grasses: deer, soft rush and sedge, which ceded in their turn to territories of moss that were then replaced by larger shrubs, gorse and bog myrtle. I spotted other smaller plants too, tormentil, blaeberry and cloudberry.

Far from being an expanse of uniformity, the moorland was constantly varying and full of colour. I came across bright yellow in the starfish leaves of butterwort, rhubarb-pink in the tentacles of sundew and snow white in the flowering heads of bog cotton. Up until the early part of the last century, this botanical diversity would have been crucially important to the lives of people in and around the Cairngorms. Previous inhabitants of the range would have been intimately familiar with the plants, both for their practical uses and superstitious significance.

Heather was used for a multitude of applications, including basket-making, thatching, bedding, for cordage and as a dye. It could even be used as a remedy for coughs as well as for making ale. Another of the most ubiquitous plants of the range has also been used for centuries. Sphagnum moss's absorbency and natural concentrations of iodine have meant it is ideal for use as a sterile dressing for wounds. It was supposedly used to staunch the injuries of Scottish soldiers after the battle of Flodden and was collected in industrial quantities from the Highlands for use during the First and Second World Wars.

Almost every species of plant would have been valued for a particular function. The roots of the flower tormentil were prized for tanning leather and as a cure for diarrhoea. Juniper was burnt in houses so that its acrid smoke would cleanse the home of pests. The shoots of soft rushes, when peeled, dried and soaked in animal fat, would serve as slow-burning candles. Bog myrtle was also highly useful as a fever-remedy, a cure for ulcers and a natural insect repellent. To this day, hill-goers often rub the crushed leaves of bog myrtle on their skin to ward off midges. Spiritual associations were also intensely valued. Clubmoss and butterwort were considered lucky and carried as charms, especially when travelling, whereas plants such as foxglove were linked with witchcraft.

Awareness of this plant lore has been all but lost in modern-day lives, but occasional fragments of this ancestral knowledge remain. They survive in contemporary customs and habits: brambles collected in autumn, mistletoe and holly brought into homes at Christmas and rowan trees planted outside houses to ward off evil spirits.

The triangulated connection between the humans, plants and mountains of the Cairngorms was keenly observed by Nan Shepherd. The 'living things', she believed, were impossible to separate from 'the forces that create them, for the mountain is one and indivisible, and rock, soil, water and air are no more integral to it than what grows from the soil and breathes air.'

Shepherd saw this interaction between plant and human life and the Cairngorms as not only mutually beneficial, but vital. 'Heather grows in its most profuse luxuriance on granite, so that the very substance of the mountain is in its life'. In doing so the roots of the plants fulfil a reciprocal purpose, binding the soil and holding fast the fabric of the mountain and the human life it sustains. This interconnectedness was central to Shepherd's unifying and radical view of the Cairngorms. It was a proposition that viewed every part of the range as 'aspects of one entity', the eponymous 'living mountain' of her short but beguiling thought-piece book.

The Living Mountain was written in the 'disturbed and uncertain world' that existed in the latter years of the Second World War and was a kind of meditative sanctuary for Shepherd. She described it as 'a secret place of ease', where she recalled memories of Cairngorm ventures many years afterwards. At the time of completion just after the war, one publisher was approached to take the book, but the manuscript was courteously rejected and spent the following thirty years stuffed in a drawer and forgotten about until its eventual publication in 1977. The book, though, is exceptional, and its concept remains ground-breaking even by contemporary standards: it is a physical, emotional and sensory character-study of a single range of mountains, a British mountain monograph quite unlike anything that has been written before or since.

Mountaineering and mountain literature have traditionally tended towards the prosaic, an evolution, partly, of the old climbing-club journals and their instructional approach to route-finding and wayfaring. In pre-modern mountain literature, where wonderment with the landscape occurs, as it did with MacGillivray and Seton Gordon, the appreciation is no less valid but is often a by-product of some other reason for being there. *The Living Mountain* is different. It is a book written in retrospect, a consigning to paper of many years of Cairngorm exploration that occurred for no other purpose than to satisfy a deep and continuing personal fascination. It was a captivation that for Shepherd, seemed inexhaustible. 'However often I walk on them,' Shepherd wrote, 'these hills hold astonishment for me. There is no getting accustomed to them.'

Shepherd's interactions with the Cairngorms were indeed vast – full-bodied connections with every aspect of their environment. She describes seeing the plateau 'glittering white . . . an immaculate vision, sun-struck, lifting against a sky of dazzling blue'. Elsewhere, she notices minute details. 'If one can look below the covering ice on a frozen burn, a lovely pattern of indentations is found, arched and chiselled, the obverse of the water's surface.'

What is so unusual about the book is that explorations are recalled that are multi-sensory, not just visual. Descriptions of the mountains include the sounds of storms, of gales crashing into corries, of cloudbursts roaring in the ravines and thunder that 'reverberates with a prolonged and menacing roll'. There are smells everywhere too, fragrances that are 'aromatic and heady', like the scent of heather and pine released by the sun, or the brandy-like perfume of birch when it rains. Elsewhere, she finds 'spicy juniper' and 'honey-sweet orchids', even the 'earthy smell of moss', the 'rank smell of deer' or the 'sharp scent of fire'. 'I draw life in through the delicate hairs of my nostrils,' Shepherd wrote.

'Touch,' Shepherd remarks, 'is the most intimate sense of all,' from the sensory pleasure she finds walking barefoot through heather to the feel of rainwater on juniper when brushed against the hand, the 'wet drops trickling over the palm', or the way cold spring water tingles the throat, wind flattens the cheeks, frost stiffens muscles and cold air makes 'lungs crackle'. Taste is another sense she evokes, in the 'subtle and sweet' flavours of cloudberries and blaeberries.

Shepherd's focus on the sensual response to her explorations of the mountains forms the basis of an almost metaphysical appreciation of the Cairngorms. Once there, she writes, 'the eye sees what it didn't see before, or sees in a new way what it had already seen. So the ear, the other senses. It is an experience that grows . . . unpredictable and unforgettable, come the hours when heaven and earth fall away and one sees a new creation'. No other British mountain range can claim such a spiritualised, philosophical discourse. The Lakes had their Romantic devotees, in particular Coleridge and Wordsworth, but nothing they wrote approaches the collective focus Shepherd brings to bear in *The Living Mountain*.

Few other writers can rival such exclusivity of concentration on one particular place, especially a mountain range. Shepherd knew the Cairngorms intimately, almost confidentially. She contoured as much as she climbed; she lingered, meandered, waited and observed, eschewing the typical linear routes to summits, searching instead for the range's 'hidden recesses'. 'Often the mountain gives itself most completely when I have no destination,' Shepherd observed, 'where I reach nowhere in particular, but have gone out merely to be with the mountains as one visits a friend.' Over time, and seen through the assemblage of her memories in *The Living Mountain*, Shepherd's journeys become a kind of lens. A way of viewing the many interlinking narratives she encounters in the range, gathered together as one indivisible form, the all-encompassing identity of the Cairngorms that she refers to ardently and collectively as 'the total mountain'.